Real-Time Embedded Systems

Optimization, Synthesis, and Networking

Real-Time Embedded Systems

Optimization, Synthesis, and Networking

Meikang Qiu
Jiayin Li

CRC Press is an imprint of the
Taylor & Francis Group, an **informa** business
A CHAPMAN & HALL BOOK

CRC Press
Taylor & Francis Group
6000 Broken Sound Parkway NW, Suite 300
Boca Raton, FL 33487-2742

First issued in paperback 2019

ISBN-13: 978-1-4398-1760-5 (hbk)
ISBN-13: 978-0-367-38267-4 (pbk)

Library of Congress Cataloging-in-Publication Data

Qiu, Meikang.
Real-time embedded systems : optimization, synthesis, and networking / Meikang Qiu, Jiayin Li.
p. cm.
Includes bibliographical references and index.
ISBN 978-1-4398-1760-5 (hardcover : alk. paper)
1. Embedded computer systems. 2. Real-time data processing. I. Li, Jiayin. II. Title.

TK7895.E42Q584 2011
006.22--dc23 2011023005

Visit the Taylor & Francis Web site at
http://www.taylorandfrancis.com

and the CRC Press Web site at
http://www.crcpress.com

Contents

Preface

Embedded systems are electronic systems that include a microprocessor to perform certain specific tasks. The microprocessor is embedded or hidden inside these products. Embedded systems are ubiquitous. Every week millions of tiny chips come out of factories like Freescale, Microchip, Philips, Texas Instruments, and Mitsubishi, finding their way into our everyday products. Engineers play a major role in all phases of this effort: planning, design, analysis, manufacturing, and marketing.

This book provides an introduction to real-time embedded systems from mainly three aspects, which represent three major trends in current embedded systems designs.

1. Optimization. This includes Chapters 2–5. In Chapter 2, we introduce the optimization in real-time embedded systems. In Chapter 3, we present scheduling algorithms in multicore embedded systems. As inaccurate information exists in the embedded systems, a robustness measurement against inaccurate information in the embedded systems is introduced in Chapter 4. We further discuss the heterogeneous optimization problem in Chapter 5.

2. Synthesis. Chapters 6–7 belong in this category. In Chapter 6, we provide a synthesis level scheduling algorithm for Phase Change Memory with Scratch Pad Memory. A thermal-aware multiprocessor synthesis technology is presented in Chapter 7.

3. Networking. This includes Chapters 8–10, which focus on task scheduling in wireless sensor network and in cloud computing, respectively. Networking and embedded systems are merging, leading to a new type of system called, *cyber physical system* (CPS).

The overall educational objective is to allow readers to discover how the computer interacts with its environment. The focus will be on understanding and analysis, with an introduction to design.

The book is focused on system level. Based on our experience, the optimization at system level has more weight than that of component level. The book is essentially organized into ten chapters. Chapter 1 is about the introduction of real-time embedded systems. Chapters 2–10 will describe the different aspects of real-time embedded systems, separately. They belong to three major parts: optimization, synthesis, and networking.

Chapters 1–3, 5–6, and 8 are written by Meikang Qiu, and Chapters 4, 7, and 9–10 are written by Jiayin Li.

Acknowledgments

We are enormously grateful to numerous individuals for their assistance in developing this book. First of all, we would like to acknowledge those who provided the excellent feedback for this book and the immeasurable help from the editors and anonymous reviewers. We also appreciate the assistance of students from our hardware/software co-design lab at the University of Kentucky.

Dr. Qiu would like to thank his wife, Diqiu Cao, son, David Qiu, father, Shiqing Qiu, mother, Longzhi Yuan, brother, Meisheng Qiu, and sister, Meitang Qiu, and the many other relatives for their continuous love, support, trust, and encouragement throughout his life. Without them, none of this would have happened.

Dr. Li would like to express his thanks to his wife, Ying, and parents, Zhihui and Huiying, for their love and support.

Authors

Meikang Qiu received B.E. and M.E. degrees from Shanghai Jiao Tong University, China. He received M.S. and Ph.D. degrees in computer science from University of Texas at Dallas in 2003 and 2007, respectively. He has worked at Chinese Helicopter R&D Institute and IBM. Currently, he is an assistant professor of ECE at the University of Kentucky, Lexington. He is an IEEE senior member and has published more than 100 peer-reviewed papers, including 35 journal articles. He has been on various chairs and served as a TPC member for many international conferences. Dr. Qiu served as the program chair of IEEE EmbeddCom09 and EM-Com09. He received the Air Force Summer Faculty Award 2009 and won the Best Paper Awards in IEEE Embedded and Ubiquitous Computing (EUC) 2009, IEEE/ACM GreenCom 2010, and IEEE CSE 2010. Dr. Qiu has three Best Paper awards and one Best Paper nomination in IEEE international conferences. He also won the Best Paper Award of ACM Transactions on Design Automation of Electronic Systems (TODAES) 2011. His research interests include embedded systems, computer security, and wireless sensor networks.

Jiayin Li received B.E. and M.E. degrees from Huazhong University of Science and Technology (HUST), China, in 2002 and 2006, respectively. Now he is pursuing his Ph.D. at the Department of Electrical and Computer Engineering (ECE), University of Kentucky. His research interests include software/hardware co-design for embedded systems and high performance computing.

Chapter 1

Introduction to Real-Time Embedded Systems

1.1 Embedded Systems Overview

Embedded systems are driving an information revolution with their pervasion in our everyday lives. These tiny, quick and smart systems can be found everywhere, ranging from commercial electronics such as cell phones, cameras, portable health monitoring systems, automobile controllers, robots, smart security devices to critical infrastructure such as telecommunication networks, electrical power grids, financial institutions, nuclear plants. The increasingly complicated embedded systems require extensive design automation and optimization tools. Architectural-level synthesis with code generation is an essential stage towards generating an embedded system satisfying stringent requirements such as time, area, reliability, and power consumption, while keeping the product cost low and development cycle short.

Recently, computing systems are everywhere. From high performance supercomputers to personal desktop computers, to mobile laptops, millions of computing systems are manufactured every year. These systems are built for general purposes. Surprisingly, more computing systems, about billions, are built for different specific purposes. These computing systems are embedded in some larger electronic device, performing certain routines of functions repeatedly. We call this computing system the embedded system.

Embedded systems include consumer electronics (digital cameras, calculators, and smart phones), home appliances (microwaves, blu-ray players, and home security systems), office and business equipment (printers, scanners, and cash registers), and automobiles (ABS, cruise control). Even though the embedded systems may not have the same computational power as the general purpose computers, they contribute a huge part in the modern world. In 2004, a family may have two or three computers, while more than 300 embedded systems exist in a normal family home.

With the advance of the technology, embedded systems with multiple cores or VLIW-like architectures, such as multicore network processors, TI's TMS320C6x, Philips' TriMedia, IBM cell processor, and IA64, etc., become necessary to achieve the required high performance for the applications with growing complexity. However, the computer-aided design capability of embedded systems has become a severe bottleneck for productivity. The design of parallel and heterogeneous embedded systems poses a host of technical challenges different from those faced by general-purpose computers because they have much larger design space and are more constrained in terms of timing, power, area, memory and other resources.

Common Characters. Embedded systems have several common characteristics that distinguish such systems from other computing systems [210, 212]:

- *Single-functioned:* An embedded system usually executes a specific program repeatedly. For example, a pager is always a pager.
- *Tightly constrained:* All computing systems have constraints on design metrics, but those on embedded systems can be especially tight. A design metric is a measure of an implementation's features, such as cost, size, performance, and power. Embedded systems often must cost just a few dollars, must be sized to fit on a single chip, must perform fast enough to process data in real-time, and must consume minimum power to extend battery life or prevent the necessity of a cooling fan.
- *Reactive and real-time:* Many embedded systems must continually react to changes in the system's environment and must compute certain results in real-time without delay. For example, a car's cruise controller continually monitors and reacts to speed and brake sensors. It must compute acceleration and deceleration amounts repeatedly within a limited time. A delayed computation could result in a failure to maintain control of the car. Basically, there are two kinds of real-time: 1) Hard real-time. There is an absolute deadline, beyond which the answer will be useless. 2) Soft real-time. The deadline missing is not catastrophic. The utility of answer will degrade with time difference from deadline.

1.1.1 Processor Technology

Processors are the computation engine in the embedded system. Compared to the processors in general purpose computing systems, the processors in embedded systems are designed for a particular function, such as signal processing. To implement the specific function, there are various approaches.

General-purpose processors — software To maximize the sale of the processor chips, the designer usually makes the processor chip programmable, so that more applications are suitable running on it. There are two key designs in this kind of processor. First of all, they need to include a general enough datapath, which is suitable for most of the applications. For this purpose, the general-purpose processors often have a large number of registers and multiple *Arithmetic Logic Units* (ALU). Second, program memory should be implemented in the memory, because in the design period, the design does not know which kind of applications would run in the processor. However, in embedded systems design, designers are more aware of how to program a specific function on general-purpose processors. They simply design the software on the general-purpose processors.

Single-purpose processors — hardware

Since designing general-purpose processors requires enormous time and effort, a single-purpose processor is more suitable for embedded systems design. Instead of designing a entire new single-purpose processor from designing a custom digital circuit, designers prefer purchasing a pre-design single-purpose processor and adapting requiring components, such as coprocessor, accelerator, and peripheral. We refer to this portion of the embedded systems design as the *hardware* portion.

Application-specific processors

An *application-specific instruction-set processor* (ASIP) is a more suitable approach in the embedded systems design. The circuit and the components in the ASIP are designed specially for the target application. Therefore, both the performance and the efficiency are optimal. The designer can optimize the processor for characteristics of the target application, such as embedded controlling, image processing, or wireless communications.

1.1.2 Design Technology

Design technology lets us convert concepts of the design into a real-world application. There are three main methods in design technology to improve the efficiency of this conversion.

Compilation and synthesis

Describing an embedded system at a high level is more productive than doing so at a low level. And compilation and synthesis can help designers in automatically generating lower level implementation details from the abstract functionality specified by the designer.

There are several different kinds of synthesis tools. The logic synthesis tool can translate the Boolean expressions into a netlist. And the *register-transfer* (RT) synthesis tool converts finite-state machines and register transfers to an RT datapath and a controller of Boolean equations. Another behavioral synthesis tool can transform a sequential program into a register. Above these, the system synthesis can convert a system specification into a chip design implementation.

Libraries and IP

Using existing libraries can also help designers in embedded systems designs. Various levels of libraries, such as logic-level libraries, RT-level libraries, and behavioral-level libraries provide layout designs in the forms of gates, RT components, controllers, and peripherals. Rather than designing every detail, designers can focus on building system architectures and optimizing, by reusing components from existing libraries. We call the layouts that can be implemented as parts of an IC as *cores*. *Intellectual properties* (IP) from the CAD industry provide a large number of cores to embedded systems designers.

Test and verification

The functionality of a system should be tested and verified before its final version. The test and verification technology can prevent the time-consuming debugging processes at the low level. Simulation is one of the most common methods for test and verification. Like other techniques mentioned above, simulations are various in three different levels, including the logic level, the RT level, and the behavioral level.

1.1.3 Memory Technology

Memory is one of the fastest evolving technologies in embedded systems over the recent decade. Following the Moore's Law, the bit-capacity of memory doubles every 18 months. No matter how fast processors can run, there is one unchanged fact that every embedded system needs memory to store data. Furthermore, due to the rapid development of the processor, more and more data are back and forth between the processor and the memory. The bandwidth of the memory, that is the speed of memory, becomes the major constraints impacting the system performance.

When building an embedded system, the designer should consider the overall performance of the memory design in the system. There are two key metrics for memory performance: *write ability* and *storage permanence*. First of all, we need to think about write ability. Writing in memory can be various in different memory technologies. Some kinds of memories, for example RAM, require special devices or techniques for writing. Otherwise, they cannot be written. Meanwhile, the write processes in different memories may need dramatically various writing time. In addition, the energy consumption is also different.

At the high end of the memory technology, we can select the memory that the processor can write to simply in a short time. There are some kinds of memories that can be accessed by setting address lines, or data bits, or control lines appropriately. At the middle of the range of memory technology, some slow written memory can be chosen. And at the low end are the types of memory that require special equipment for writing.

Meanwhile, we also need to consider storage permanence. How long the memory can hold the written bits in themselves can have a key impact on the reliability of the system. In the aspect of storage permanence, there are two kinds of memory technologies: *nonvolatile* and *volatile*. The major difference is that the nonvolatile memory can hold the written bits after power is no longer supplied, but volatile cannot.

1.1.4 Design Challenge–Optimizing Design Metrics

The embedded systems designer must of course construct an implementation that fulfills desired functionality, but a difficult challenge is to construct an implementation that simultaneously optimizes numerous design metrics.

Metrics typically compete with one another: Improving one often leads to worsening of another. For example, if we reduce an implementation's size, the implementation's performance may suffer. Some observers have compared this phenomenon to a wheel with numerous pins. If you push one pin in, such as size, then the other pins pop out. To best meet this optimization challenge, the designer must be comfortable with a variety of hardware and software implementation technologies, and must be able to migrate from one technology to another, in order to find the best implementation for a given application and constraints [210, 212].

- **Time-to-Market Design Metric:** The time-to-market constraint has become especially demanding in recent years. Introducing an embedded system to the marketplace early can make a big difference in the system's profitability.

- **Performance:** The execution time of the systems. The two main measures of performance are: latency and throughput.

- **Low Cost:** Cost is very sensitive for embedded systems. Cost includes nonrecurring cost (one-time monetary cost) and unit cost (monetary cost of each copy of the system).

- **Small Size:** Physical size is an important design metric. For example, handheld electronics need to be small. A brake system in a car cannot be too large.

- **Low Weight:** One of the biggest concerns for handheld device is weight. Also, for transportation applications, weight will cost money.

- **Low Power:** Battery power capacity has limited increase in recent years. For embedded systems, limited cooling may limit power even if AC power is available.

- **Adaptivity:** Embedded systems are usually worked under harsh environments with: a) heat, vibration, and shock; b) power fluctuations, RF interference, and lightning; and c) water, corrosion, and physical abuse. The ability to deal with this is critical to the applications.

- **Safety:** This means embedded systems must function correctly and must not function incorrectly.

1.2 Fixed Time Model of Heterogeneous Embedded Systems

Heterogeneity is one of the most important features of modern embedded systems. We model the *heterogeneous embedded systems* (HES) as two types according to the execution time scenario. If the execution time of each task is a fixed value, we call it a *fixed time model.* Otherwise, if the execution time is a non-fixed value, we model it as a random variable and call this type *probabilistic time model.*

With more and more different types of functional units (FUs) available, the same type of operations can be processed by heterogeneous FUs with different costs, where the cost may relate to power, reliability, or monetary cost. We want to design efficient techniques that can assign a proper FU type to each operation of a DSP application and generate a schedule in such a way that the requirements can be met and the total cost can be minimized.

We propose [183] a two-phase approach to solve this problem. In the first phase, we solve hard HA (*heterogeneous assignment*) problems, i.e., given the types of heterogeneous FUs, a *Data Flow Graph* (DFG) in which each node has different execution times and costs for different FU types, and a timing constraint, how to assign a proper FU type to each node such that the total cost can be minimized while the timing constraint is satisfied. In the second phase, based on the assignments obtained in the first phase, we propose a *minimum resource scheduling algorithm* to generate a schedule and a feasible configuration that uses as little resource as possible. Here, a configuration means which FU types and how many FUs for each type should be selected in a system.

An exemplary DFG is shown in Figure 1.1(a). Assume we can select FUs from a FU library that provides three types of FUs: P_1, P_2, P_3. The execution times and costs of each node for different FU types are shown in Figure 1.1(b). In Figure 1.1(b), column "T_i" presents the execution time, and column "C_i" presents the execution cost for each FU type P_i.

The execution cost can be any cost such as energy consumption or reliability cost. A node may run slower but with less energy consumption or reliability cost when executed on one type of FUs than on another. When the cost is related to energy consumption, it is clear that the total energy consumption is the summation of energy cost of each node. Also, when the execution cost is related to reliability, the total reliability cost is the summation of reliability cost of all nodes. We compute the reliability cost using the same model as in [196].

Define the reliability of a system as the probability that the system will not fail during the time of executing a DFG. Consider a heterogeneous system with M FU types, $\{P_1, P_2, \cdots, P_M\}$, and a DFG containing N nodes, $\{u_1, u_2, \cdots, u_N\}$. Let $t_j(i)$ be the execution time of node u_i for type P_j. Let

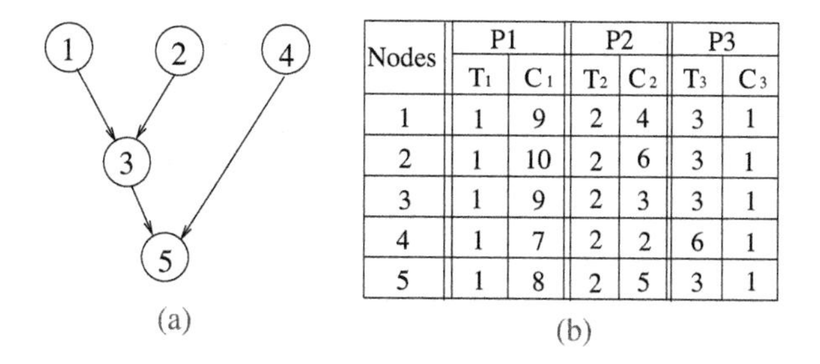

Nodes	P1		P2		P3	
	T_1	C_1	T_2	C_2	T_3	C_3
1	1	9	2	4	3	1
2	1	10	2	6	3	1
3	1	9	2	3	3	1
4	1	7	2	2	6	1
5	1	8	2	5	3	1

(b)

FIGURE 1.1: A given DFG and the execution times and costs of its nodes for different FU types.

f_j be the failure rate of type P_j. Then the *reliability cost* of node u_i for type P_j is defined as $t_j(i) \times f_j$. In order to increase the reliability of a system, we need to reduce the summation of reliability costs of all nodes. The reason is as follows. Let x_{ij} be a binary number that denotes whether type P_j is assigned to node u_i or not (it equals 1 if P_j is assigned to u_i; otherwise it equals 0). The probability of a system not to fail during the time of processing a DFG, is:

$$Pr = \prod_{1 \le j \le M, 1 \le i \le N} (1 - f_j)^{x_{ij} t_j(i)}.$$

From this equation, we know $Pr \approx \prod(e^{-f_j x_{ij} t_j(i)})$ when f_j is small [185].

An exemplary DFG is shown in Figure 1.1(a). Thus, in order to maximize Pr, we need to minimize $\sum(f_j x_{ij} t_j(i))$. In other words, we need to find an assignment such that the timing constraint is satisfied and the summation of reliability costs of all nodes is minimized in order to maximize the reliability of a system.

Nodes	P1	P2	P3
1		✓ (4)	
2		✓ (6)	
3		✓ (3)	
4		✓ (2)	
5		✓ (5)	

(a)

Nodes	P1	P2	P3
1			✓ (1)
2			✓ (1)
3		✓ (3)	
4		✓ (2)	
5	✓ (8)		

(b)

FIGURE 1.2: (a) Assignment 1 with cost 20. (b) Assignment 2 with cost 15.

Assume the given costs are energy consumption and the timing constraint is 6 time units in this example. For the given DFG in Figure 1.1(a) and the time

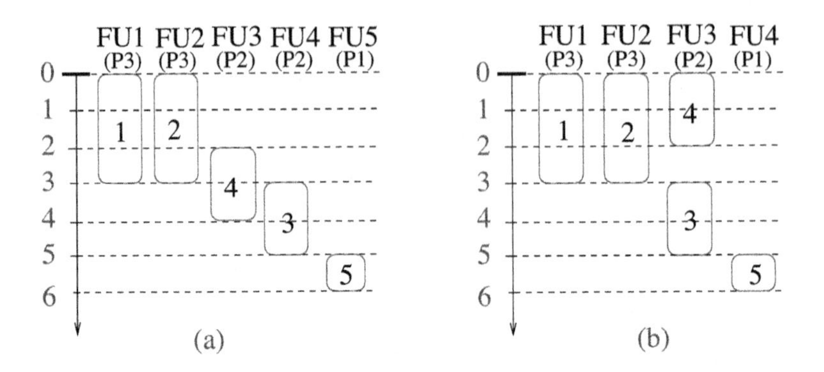

FIGURE 1.3: Two schedules corresponding to Assignment 2.

cost table in Figure 1.1(b), two different assignments are shown in Figure 1.2. In Figure 1.2, if a FU type is assigned to a node, "$\surd$" is put into the right location and the value in the parentheses beside "$\surd$" is the corresponding execution cost. The total execution cost for Assignment 1 is 20. The total cost for Assignment 2 is 15 and this is an optimal solution, which is 25% less than that for Assignment 1. Our assignment algorithm in [183] achieves the optimal solution for this example.

For Assignment 2, two different schedules with corresponding configuration are shown in Figure 1.3. The configuration in Figure 1.3(a) uses five FUs while the configuration in Figure 1.3(b) uses four FUs. The schedule in Figure 1.3(b) is generated by the minimum resource scheduling algorithm in [183], in which the configuration achieves the minimal resource for Assignment 2.

To solve the heterogeneous assignment problem, when the given DFG is a simple path or a tree, our algorithms can produce the optimal solutions [183, 160, 164]. These algorithms are efficient in practice, though rigorously speaking they are pseudo polynomial because the complexities are related to the value of the maximum execution time of nodes. But this value is usually not large or can be normalized to be small. Then, based on the obtained assignment, a minimum resource scheduling algorithm is proposed to generate a schedule and a configuration. The experimental results [183] show that our algorithms on average give a reduction of 27.4% on system cost compared with the other existing methods.

1.3 Probabilistic Time Model of Heterogeneous Embedded Systems

We are continuing our work to deal with a much tougher model: probabilistic time model. In many embedded applications, some tasks may not have fixed execution times. Such tasks usually contain conditional instructions and/or operations that could have different execution times for different inputs. Existing methods are not able to deal with such uncertainties, so either worst-case or average-case computation times for these tasks are usually assumed. Such assumptions, however, may result in an inefficient design, especially for soft real-time systems.

In papers [160, 164, 161, 165], we have studied the optimization algorithms which operate in probabilistic environments to solve the *heterogeneous assignment with probability* (HAP) problem. In the HAP problem, we model the execution time of a task as a random variable [241]. For heterogeneous systems, each FU is associated with a cost related to hardware cost, area, reliability, power consumption, etc. Faster one has higher cost while slower one has lower cost. We need to study how to assign a proper FU type to each node of a DFG such that the total cost is minimized while the timing constraint is satisfied with a guaranteed confidence probability. With a confidence probability P, we can guarantee that the total execution time of the DFG is less than or equal to the timing constraint with a probability greater than or equal to P.

Here we show an example to illustrate the HAP problem. Assume that we can select FUs from a FU library that provides two types of FUs: R_1, R_2. An exemplary PDFG (Probabilistic Data Flow Graph) is shown in Figure 1.4(a), which is a tree with four nodes. The execution times (T), probabilities (P), and costs (C) of each node for different FU types are shown in Figure 1.4(b). Each node can choose one of two FU types, and executes each type of FU with probabilistic execution times. Thus, execution time (T) of each FU is a random variable. For example, when choosing R_1, node 1 can be executed in 1 time unit with probability 0.9 and be executed in 3 time units with probability of 0.1. In other words, node 1 can be finished in 3 time units with 100% probability.

We have studied [160, 164, 161, 165] to find efficient solutions of the HAP problem. Note that when only considering worst-case times, the designed algorithms will solve the hard HA problem. Hence, our algorithms can produce solutions for both hard and soft real-time situations. Second, the proposed algorithms can provide more choices with guaranteed confidence probabilities and smaller total costs. Yet, the *optimal hard HA algorithms* may not find solutions with certain timing constraints. Preliminary experiments show that, when the input PDFG is a tree or a simple path, the result of probabilistic approach has an average 32.5% improvement with 0.9 confidence probability compared with the best results to the hard HA problem. We are working on generalizing our algorithms to any data flow graphs.

The development of the probabilistic heterogeneous scheduling and assignment framework can be easily applied to the solution of dynamic voltage scaling problem for probabilistic graph models. As the models presented in Hua et al. [92, 91, 90], each node can be associated with different execution time and energy consumption with probabilities; the problem is to find the appropriate voltage levels for each node such that the total energy can be minimized. In this project, we will further attack this problem by using loop scheduling with multicore architecture.

By using the HAP framework, we can regard each voltage level as a FU type and each expected energy as the cost. Then this problem becomes a special case of the HAP problem. In this project, we will focus on the time and power optimization. The success of this research will give us a stimulation to generate an optimization framework for a heterogeneous embedded system. We believe that the development of the HAP framework will have a significant impact because many optimization problems should be intrinsically modeled as probabilistic models such as cache/memory access, timing, reliability, etc. The algorithms should directly work on the probabilistic distributions, instead of operating on fixed values assumed by worst cases or expected cases.

1.3.1 Background and Example

This section presents assignment and optimization algorithms which operate in probabilistic environments to solve the *heterogeneous assignment with probability* (HAP) problem. In the HAP problem, we model the execution time of a task as a random variable. For heterogeneous systems, each FU has a different cost, representing hardware cost, size, reliability, etc.. Faster one has higher cost while slower one has lower cost. This chapter shows how to assign a proper FU type to each node of a *Data Flow Graph* (DFG) such that the total cost is minimized while the timing constraint is satisfied with a guaranteed confidence probability. With confidence probability P, we can guarantee that the total execution time of the DFG is less than or equal to the timing constraint with a probability that is greater than or equal to P.

We show an example to illustrate the HAP problem. Assume that we can select FUs from a FU library that provides two types of FUs: R_1, R_2. An exemplary DFG is shown in Figure 1.4(a), which is a *directed acyclic graph* (DAG) with 4 nodes. Node 0 is a multi-child node, which has two children: 1 and 2. Node 3 is a multi-parent node, and has two parents: 1 and 2. The execution times (T), probabilities (P), and costs (C) of each node for different FU types are shown in Figure 1.4(b). Each node can select one of the two different FU types. The execution time (T) of each FU type is modeled as a random variable; and the probabilities may come from the statistical profiling. For example, node 0 can choose one of the two types: R_1 or R_2. When choosing R_1, node 0 will be finished in 1 time unit with probability 0.9

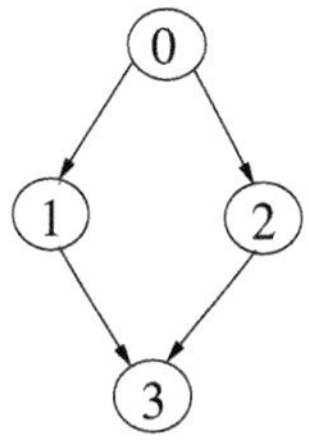

(a)

Nodes	R1			R2		
	T	P	C	T	P	C
0	1	0.9	10	2	0.7	4
	3	0.1	10	4	0.3	4
1	1	0.8	8	3	0.6	3
	4	0.2	8	5	0.4	3
2	1	0.8	6	3	0.6	2
	4	0.2	6	5	0.4	2
3	1	0.9	9	2	0.8	5
	4	0.1	9	6	0.2	5

(b)

Nodes	R1			R2		
	T	P	C	T	P	C
0	1	0.9	10	2	0.7	4
	3	1.0	10	4	1.0	4
1	1	0.8	8	3	0.6	3
	4	1.0	8	5	1.0	3
2	1	0.8	6	3	0.6	2
	4	1.0	6	5	1.0	2
3	1	0.9	9	2	0.8	5
	4	1.0	9	6	1.0	5

(c)

FIGURE 1.4: (a) A given DAG. (b) The times, probabilities, and costs of its nodes for different FU types. (c) The time cumulative distribution functions (CDFs) and costs of its nodes for different FU types.

and will be finished in 3 time units with probability 0.1. In other words, node 0 can guarantee to be finished in 3 time units with 100% probability. Hence, we care about the time *cumulative distribution function* (CDF) $F(t)$, which gives accumulated probability for $T \leq t$. Figure 1.4(c) shows the time CDFs and costs of each node for different FU types.

A solution to the HAP problem with timing constraint 11 can be found as follows: We assign FU types 2, 2, 2, and 1 for nodes 0, 1, 2, and 3, respectively. Let T_0, T_1, T_2, and T_3 be the random variables representing the execution times of nodes 0, 1, 2, and 3. From Figure 1.4(c), we get: $Pr(T_0 \leq 4) = 1.0$, $Pr(T_1 \leq 5) = 1.0$, $Pr(T_2 \leq 5) = 1.0$, and $Pr(T_3 \leq 1) = 0.9$. Hence, we obtain minimum total cost 18 with 0.9 probability satisfying the timing constraint 11. The total cost is computed by adding the costs of all nodes together and the probability corresponding to the total cost is computed by multiplying the probabilities of all nodes.

In Figure 1.4(b), if we use the worst-case execution time as a fixed execution time for each node, then the assignment problem becomes the *hard heterogeneous assignment* (hard HA) problem, which is related to the hard real-time. The hard HA problem is the worst-case scenario of the *heterogeneous assignment with probability* (HAP) problem, and is easy to compute. For example, in the hard HA problem, when choosing type R_1, node 0 has only one execution time 3. When choosing type R_2, node 0 has one execution time 4. With certain timing constraints, there might not be a solution for the hard HA problem. However, for soft real-time applications, it is desirable to find an assignment that guarantees the total execution time to be less than or equal to the timing constraint with certain confidence probability. Solving HAP problem is a complicated task, but the result we achieved deserves the efforts we spend for it.

For example, in Figure 1.4, under timing constraint 11, we cannot find a solution to the hard HA problem. But we can obtain minimum system cost 18 with probability 0.9 satisfying the timing constraint 11. Also, the cost obtained from the worst-case scenario is always larger than or equal to the cost from the probabilistic scenario. For example, under timing constraint 11, the minimum cost is 33 for the hard HA problem. While in the HAP problem, with confidence probability 0.9 satisfying the timing constraint, we get the minimum cost of 18, which gives 45.5% improvement.

1.3.1.1 System Model

We use *Data-Flow Graph (DFG)* to model a DSP application. A ***DFG G*** $= \langle V, E \rangle$ is a *directed acyclic graph* (DAG), where $V = \langle v_1, v_2, \cdots, v_N \rangle$ is the set of nodes, $E \subseteq V \times V$ is the edge set that defines the precedence relations among nodes in V.

In practice, many architectures consist of different types of FUs. Assume there are maximum M different FU types in a FU set R=$\{R_1, R_2, \cdots, R_M\}$. An assignment for a DFG G is to assign a FU type to each node. Define an ***Assignment A*** to be a function from domain V to range R, where V is the node set and R is FU type set. For a node $v \in V$, $A(v)$ gives the selected type of node v. For example, in Figure 1.4(a), Assigning FU types 2, 2, 2, and 1 for nodes 0, 1, 2, and 3, respectively, we obtain minimum total cost 18 with 0.9 probability satisfying the timing constraint 11. That is, $A(0) = 2$, $A(1) = 2$, $A(2) = 2$, and $A(3) = 1$.

In a DFG G, each varied execution time is modeled as a probabilistic random variable. $\mathbf{T_{R_j}(v)}$ $(1 \le j \le M)$ represents the execution times of each node $v \in V$ for FU type j, and $\mathbf{P_{R_j}(v)}$ $(1 \le j \le M)$ represents the corresponding probability function. And $\mathbf{C_{R_j}(v)}$ $(1 \le j \le M)$ is used to represent the cost of each node $v \in V$ for FU type j, which is a fixed value. For instance, in Figure 1.4(a), $T_1(0) = 1, 3$; $T_2(0) = 2, 4$. Correspondingly, $P_1(0) = 0.9, 0.1$; $P_2(0) = 0.7, 0.3$. And $C_1(0) = 10$; $C_2(0) = 4$.

We define the ***heterogeneous assignment with probability (HAP)*** problem as follows: Given M different FU types: R_1,R_2,$\cdots$,R_M, a DFG $G = \langle V, E \rangle$ where V=$\langle v_1$,v_2,$\cdots$,$v_N \rangle$, $T_{R_j}(v)$, $P_{R_j}(v)$, $C_{R_j}(v)$ for each node $v \in V$ executed on each FU type j, and a timing constraint L, find an assignment for G that gives the *minimum total cost C with confidence probability P under timing constraint L*. In Figure 1.4(a), a solution to the HAP problem under timing constraint 11 can be found as follows. Assigning FU types 2, 2, 2, and 1 for nodes 0, 1, 2, and 3, respectively, we obtain minimum total cost 18 with 0.9 probability under the timing constraint 11.

1.3.1.2 Example

In this subsection, we continue the example of Figure 1.4 to illustrate the *heterogeneous assignment probability* (HAP) problem. In Figure 1.4(a), each node has two different FU types to choose from, and is executed on them with probabilistic times. In DSP applications, a real-time system does not always have hard deadline time. The execution time can be smaller than the hard deadline time with certain probabilities. So the hard deadline time is the worst case of the varied smaller time cases. If we consider these time variations, we can achieve a better minimum cost with satisfying confidence probabilities under timing constraints.

T	(P , C)	(P , C)	(P , C)	(P , C)	(P , C)
3	0.52, 33				
4	0.40, 27	0.46, 29	0.52, 33		
5	0.36, 23	0.40, 27	0.46, 29	0.58, 33	
6	0.26, 20	0.30, 22	0.36, 23	0.58, 27	0.81 33
7	0.20, 14	0.27, 18	0.38, 22	0.51, 23	0.81, 24
8	0.20, 14	0.63, 18	0.72, 20	0.81, 24	0.90, 33
9	0.56, 14	0.63, 18	0.72, 20	0.90, 24	
10	0.56, 14	0.90, 18			
11	0.80, 14	0.90, 18	**1.00, 33**		
12	0.80, 14	0.90, 18	**1.00, 24**		
13	0.80, 14	**1.00, 18**			
14	0.80, 14	**1.00, 20**			
15	**1.00, 14**				

TABLE 1.1: Minimum total costs with computed confidence probabilities under various timing constraints for a DAG.

For this DAG with four nodes, we can obtain the minimum total cost with computed confidence probabilities under various timing constraints. The results generated by our algorithms are shown in Table 1.1. The entries with probability that is equal to 1 (see the entries in boldface) actually give the results to the hard HA problem which show the worst-case scenario of the HAP problem. For each row of the table, the C in each (P, C) pair gives the minimum total cost with confidence probability P under timing constraint T. For example, when $T = 3$, with pair (0.52, 33), we can achieve minimum total cost 33 with confidence probability 0.52 under timing constraint 3.

Assign	0	2	4	1.0000	4
Assign	1	2	5	1.0000	3
Assign	2	2	5	1.0000	2
Assign	3	1	1	0.9000	9
Total			18	0.9000	18
Assign	0	1	3	1.0000	10
Assign	1	1	4	1.0000	8
Assign	2	1	4	1.0000	6
Assign	3	1	4	1.0000	9
Total			11	1.0000	33

TABLE 1.2: With timing constraint 11, the assignments of types for each node with different (Probability, Cost) pairs.

Comparing with optimal results of the hard HA problem for a DFG using worst-case scenario, Table 1.1 provides more information and more selections, no matter whether the system is hard or soft real-time. In Table 1.1, we have the output of our algorithm from timing constraint 3 to 15, while the optimal results of the hard HA problem for a DAG only has five entries (in boldface) from timing constraint 11 to 15.

For a soft real-time system, some nodes of DFG have smaller probabilistic execution times compared with the hard deadline time. We can achieve much smaller cost than the cost of worst case with guaranteed confidence probability. For example, under timing constraint 11, we can select the pair (0.90, 18) which guarantees to achieve minimum cost 18 with 90% confidence satisfying the timing constraint 11. It achieves 45.5% reduction in cost compared with the cost 33, the result obtained by the algorithms using worst-case scenario of the HAP problem. In many situations, this assignment is good enough for users. We also can select the pair (0.80, 14) which has provable confidence probability 0.8 satisfying the timing constraint 11, while the cost 14 is only 42.4% of the cost of worst case, 33. The assignments for each pair of (0.9, 18) and (1.0, 33) under the timing constraint 11 are shown in Table 1.2.

From the above example, we can see that the probabilistic approach to the heterogeneous assignment problem has great advantages: It provides the possibility to reduce the total cost of systems with guaranteed confidence probabilities under different timing constraints. It is suitable to both hard and soft real-time systems. We will give related algorithms and experiments in later sections.

1.3.2 The Algorithms for the HAP Problem

In this section, we propose two algorithms to solve the general case of the HAP problem, that is, the given input is a *directed acyclic graph* (DAG).

1.3.2.1 Definitions and Lemma

To solve the HAP problem, we use a dynamic programming method traveling the graph in bottom up fashion. For the easiness of explanation, we will index the nodes based on bottom up sequence. Define a *root* node to be a node without any parent and a *leaf* node to be a node without any child. In multi-child case, we use bottom up approach; and in multi-parent case, we use top down approach.

Given the timing constraint L, a DFG G, and an assignment A, we first give several definitions as follows: 1) $\mathbf{G^i}$: The sub-graph rooted at node v_i, containing all the nodes reached by node v_i. In our algorithm, each step will add one node which becomes the root of its sub-graph. 2) $\mathbf{C_A(G^i)}$ and $\mathbf{T_A(G^i)}$: The total cost and total execution time of G^i under the assignment A. In our algorithm, each step will achieve the minimum total cost of G^i with computed confidence probabilities under various timing constraints. 3) In our algorithm, table $D_{i,j}$ will be built. Each entry of table $D_{i,j}$ will store a link list of (Probability, Cost) pairs sorted by probability in ascending order. Here we define the **(Probability, Cost) pair** $(\mathbf{C_{i,j}}, \mathbf{P_{i,j}})$ as follows: $C_{i,j}$ is the minimum cost of $C_A(G^i)$ computed by all assignments A satisfying $T_A(G^i) \leq j$ with probability $\geq P_{i,j}$.

We introduce the ***operator*** "$\oplus$" in this chapter. For two (Probability, Cost) pairs H_1 and H_2, if H_1 is $(P_{i,j}^1, C_{i,j}^1)$, and H_2 is $(P_{i,j}^2, C_{i,j}^2)$, then, after the $\oplus$ operation between H_1 and H_2, we get pair (P', C'), where $P' = P_{i,j}^1 * P_{i,j}^2$ and $C' = C_{i,j}^1 + C_{i,j}^2$. We denote this operation as "$\mathbf{H_1 \oplus H_2}$".

In our algorithm, $D_{i,j}$ is the table in which each entry has a link list that stores pair $(P_{i,j}, C_{i,j})$. Here, i represents a node number, and j represents time. For example, a link list can be (0.1, 2)→(0.3, 3)→(0.8, 6)→(1.0, 12). Usually, there are redundant pairs in a link list. We give the redundant-pair removal algorithm Algorithm 1.3.1.

Algorithm 1.3.1 Redundant-Pair Removal Algorithm

Require: A list of $(P_{i,j}^k, C_{i,j}^k)$
Ensure: A redundant-pair free list

```
1: Sort the list by P_{i,j} in an ascending order such that P^k_{i,j} ≤ P^{k+1}_{i,j}.
2: From the beginning to the end of the list,
3: for each two neighboring pairs (P^k_{i,j}, C^k_{i,j}) and (P^{k+1}_{i,j}, C^{k+1}_{i,j}) do
4:    if P^k_{i,j} = P^{k+1}_{i,j} then
5:       if C^k_{i,j} ≥ C^{k+1}_{i,j} then
6:          cancel the pair P^k_{i,j}, C^k_{i,j}
7:       else
8:          cancel the pair P^{k+1}_{i,j}, C^{k+1}_{i,j}
9:       end if
10:   else
11:      if P^k_{i,j} ≥ P^{k+1}_{i,j} then
12:         cancel the pair (P^{k+1}_{i,j}, C^{k+1}_{i,j})
13:      end if
14:   end if
15: end for
```

For example, we have a list with pairs (0.1, 2) → (0.3, 3) → (0.5, 3) → (0.3, 4), we do the redundant-pair removal as follow: First, sort the list according $P_{i,j}$ in an ascending order. This list becomes to (0.1, 2) → (0.3, 3) → (0.3, 4) → (0.5, 3). Second, cancel redundant pairs. Comparing (0.1, 2) and (0.3, 3), we keep both. For the two pairs (0.3, 3) and (0.3, 4), we cancel pair (0.3, 4) since the cost 4 is bigger than 3 in pair (0.3, 3). Comparing (0.3, 3) and (0.5, 3), we cancel (0.3, 3) since $0.3 < 0.5$ while $3 \geq 3$. There is no information lost in redundant-pair removal.

In summary, we can use the following Lemma to cancel redundant pairs.

Lemma 1.3.1 *Given $(P_{i,j}^1, C_{i,j}^1)$ and $(P_{i,j}^2, C_{i,j}^2)$ in the same list:*

1. *If $P_{i,j}^1 = P_{i,j}^2$, then the pair with minimum $C_{i,j}$ is selected to be kept.*

2. *If $P_{i,j}^1 < P_{i,j}^2$ and $C_{i,j}^1 \geq C_{i,j}^2$, then $C_{i,j}^2$ is selected to be kept.*

Using Lemma 1.3.1, we can cancel many redundant-pair ($P_{i,j}$, $C_{i,j}$) whenever we find conflicting pairs in a list during a computation. After the $\oplus$ operation and redundant pair removal, the list of ($P_{i,j}$, $C_{i,j}$) has the following properties:

Lemma 1.3.2 *For any ($P_{i,j}^1$, $C_{i,j}^1$) and ($P_{i,j}^2$, $C_{i,j}^2$) in the same list:*

1. $P_{i,j}^1 \neq P_{i,j}^2$ *and* $C_{i,j}^1 \neq C_{i,j}^2$.
2. $P_{i,j}^1 < P_{i,j}^2$ *if and only if* $C_{i,j}^1 < C_{i,j}^2$.

Since the link list is in an ascending order by probabilities, if $P_{i,j}^1 < P_{i,j}^2$, while $C_{i,j}^1 \geq C_{i,j}^2$, based on the definitions, we can guarantee with $P_{i,j}^1$ to find smaller energy consumption $C_{i,j}^2$. Hence ($P_{i,j}^2$, $C_{i,j}^2$) has already covered ($P_{i,j}^1$, $C_{i,j}^1$). We can cancel the pair ($P_{i,j}^1$, $C_{i,j}^1$). For example, we have two pairs: (0.1, 3) and (0.6, 2). Since (0.6, 2) is better than (0.1, 3), in other words, (0.6, 2) covers (0.1, 3), we can cancel (0.1, 3) and will not lose useful information. We can prove the vice versa is true similarly. When $P_{i,j}^1 = P_{i,j}^2$, smaller C is selected.

In every step in our algorithm, one more node will be included for consideration. The information of this node is stored in local table $E_{i,j}$, which is similar to table $D_{i,j}$. A local table only stores information, such as probabilities and costs, of a node itself. Table $E_{i,j}$ is the local table only storing the information of node v_i. In more detail, $E_{i,j}$ is a local table of link lists that store pair ($p_{i,j}$, $c_{i,j}$) sorted by $p_{i,j}$ in an ascending order; $c_{i,j}$ is the cost only for node v_i at time j, and $p_{i,j}$ is the corresponding probability. The building procedures of $E_{i,j}$ are as follows. First, sort the execution time variations in an ascending order. Then, accumulate the probabilities of same type. Finally, let $L_{i,j}$ be the link list in each entry of $E_{i,j}$, insert $L_{i,j}$ into $L_{i,j+1}$ while redundant pairs cancel out based on Lemma 1.3.1. For example, node 0 in Figure 1.4(b) has the following (T: P, C) pairs: (1: 0.9, 10), (3: 0.1, 10) for type R_1, and (2: 0.7, 4), (4: 0.3, 4) for type R_2. After sorting and accumulating, we get (1: 0.9, 10), (2: 0.7, 4), (3: 1.0, 10), and (4: 1.0, 4). We obtain Table 1.3 after the insertion.

Similarly, for two link lists L_1 and L_2, the operation "$\mathbf{L_1} \oplus \mathbf{L_2}$" is implemented as follows: First, implement $\oplus$ operation on all possible combinations of two pairs from different link lists. Then insert the new pairs into a new link list and remove redundant-pair using Lemma 1.3.1.

Time	1	2	3	4
(P_i, C_i)	(0.9, 10)	(0.7, 4) (0.9, 10)	(0.7, 4) (1.0, 10)	(1.0, 4)

TABLE 1.3: An example of local table, $E_{0,j}$.

1.3.2.2 A Heuristic Algorithm for DAG

The general problem of the HAP problem is NP-complete problem. We propose a near optimal heuristic algorithm in this subsection and an optimal one in next subsection. In many cases, the near optimal heuristic algorithm will give us the same results as the results of the optimal algorithm. The optimal algorithm is suitable for the cases when the given DFG has a small number of multi-parent and multi-child nodes.

The DAG_Heu Algorithm

Require: M different types of FUs, a DAG, and the timing constraint L
Ensure: An heuristic assignment for the DAG

1. Topological sort all the nodes, and get a sequence A.
   Count the number of multi-parent nodes p and the number of multi-child nodes q.
   if $p < q$ **then**
   use bottom up approach
   else
   use top down approach
   end if

2. For bottom up approach, use the following algorithm. For top down approach, just reverse the sequence.
   Assume the sequence after topological sorting is $v_1 \to v_2 \to \cdots \to v_N$, in bottom up fashion. Let $D_{1,j} = E_{1,j}$. Assume $D'_{i,j}$ is the table that stored minimum total cost with computed confidence probabilities under the timing constraint j for the sub-graph rooted on v_i except v_i. Nodes $v_{i_1}, v_{i_2}, \cdots, v_{i_R}$ are all child nodes of node v_i and R is the number of child nodes of node v_i, then

$$D'_{i,j} = \begin{cases} (0,0) & \text{if } R = 0 \\ D_{i_1,j} & \text{if } R = 1 \\ D_{i_1,j} \oplus \cdots \oplus D_{i_R,j} & \text{if } R \geq 1 \end{cases} \tag{1.1}$$

   Then, for each k in $E_{i,k}$

$$D_{i,j} = D'_{i,j-k} \oplus E_{i,k} \tag{1.2}$$

   //In equation (1.1), $D_{i_1,j} \oplus D_{i_2,j}$ is computed as follows. Let G' be the union of all nodes in the graphs rooted at nodes v_{i_1} and v_{i_2}. Travel all the graphs rooted at nodes v_{i_1} and v_{i_2}. If a node a in G' appears for the first time, we add the cost of a and multiply the probability of a to $D'_{i,j}$. If a appears more than once, that is, a is a common node, then use a selection function to choose the type of a. For instance, the selection function can be defined as selecting the type that has a smaller execution time. Therefore, we can compute $D'_{i,j}$ by using "$\oplus$" on all child nodes of node v_i when $R \geq 1$. //

3. Return $D_{N,j}$.

In algorithm *DAG_Heu*, if using bottom up approach, for each sequence node, use the simple path algorithm to get the dynamic table of parent node. If using top down approach, reverse the sequence and use the same algorithm. For example, in Figure 1.4(a), there is one multi-child node: node 0; and there is also one multi-parent node that is, node 3. Hence, we can use either approach.

In algorithm *DAG_Heu*, we have to solve the problem of common nodes, that is, one node appears in two or more graphs that are rooted at the child nodes of node v_i. In Equation (1.1), even if there are common nodes, we must not count the same node twice. That is, the cost is just added once, and the probability is multiplied once. If a common node has conflicting FU type selection, then we need to define a selection function to decide which FU type should be chosen for the common node. For example, we can select the FU type as the type that has a smaller execution time.

Due to the problem of common nodes, algorithm *DAG_Heu* is not an optimal algorithm. The reason is that an assignment conflict for a common node may exist, while algorithm *DAG_Heu* cannot solve this problem. For example, in Figure 1.4, using bottom up approach, there will be two paths from node 3 to node 0. Path a is $3 \rightarrow 1 \rightarrow 0$, and path b is $3 \rightarrow 2 \rightarrow 0$. Hence, node 3 is a common node for both paths while node 0 is the root. It is possible that, under a timing constraint, the best assignment for path a gives node 3 assignment as FU type 1, while the best assignment for path b gives node 3 assignment as FU type 2. This kind of assignment conflict cannot be solved by algorithm *DAG_Heu*. Hence, *DAG_Heu* is not an optimal algorithm, although it is very efficient in practice.

From algorithm *DAG_Heu*, we know $D_{N,L}$ records the minimum total cost of the whole path within the timing constraint L. We can record the corresponding FU type assignment of each node when computing the minimum system cost in step 2 of the algorithm *DAG_heu*. Using this information, we can get an optimal assignment by tracing how to reach $D_{N,L}$.

It takes at most $O(|V|)$ to compute common nodes for each node in the algorithm *DAG_heu*, where $|V|$ is the number of nodes. Thus, the complexity of the algorithm *DAG_heu* is $O(|V|^2 * L * M * K)$, where L is the given timing constraint. Usually, the execution time of each node is upper bounded by a constant. Then L equals $O(|V|^c)$ (c is a constant). In this case, *DAG_heu* is polynomial.

1.3.2.3 An Optimal Algorithm for DAG

In this subsection, we give the optimal algorithm (*DAG_Opt*) for the HAP problem when the given DFG is a DAG. In *DAG_Opt*, we exhaust all the possible assignments of multi-parent or multi-child nodes. Without loss of generality, assume we are using bottom up approach. If the total number of nodes with multi-parent is t, and there are maximum K variations for the execution times of all nodes, then we will give each of these t nodes a fixed assignment. We will exhaust all of the K^t possible fixed assignments by algorithm *DAG_Heu* without using the selection function since there is no assignment conflict for a common node.

The DAG_Opt Algorithm

Require: M different types of FUs, a DAG, and the timing constraint L
Ensure: An optimal assignment for the DAG

1. Topological sort all the nodes, and get a sequence A.
   Count the number of multi-parent nodes p and the number of multi-child nodes q.
   if $p < q$ **then**
     use bottom up approach
   else
     use top down approach
   end if

2. For bottom up approach, use the following algorithm. For top down approach, just reverse the sequence. If the total number of nodes with multi-parent is t, and there are maximum K variations for the execution times of all nodes, then we will give each of these t nodes a fixed assignment. For each of the K^t possible fixed assignments, we do as follows.

3. Assume the sequence after topological sorting is $v_1 \rightarrow v_2 \rightarrow \cdots \rightarrow v_N$, in bottom up fashion. Let $D_{1,j} = E_{1,j}$. Assume $D'_{i,j}$ is the table that stored minimum total cost with computed confidence probabilities under the timing constraint j for the sub-graph rooted on v_i except v_i. Nodes $v_{i_1}, v_{i_2}, \cdots, v_{i_R}$ are all child nodes of node v_i and R is the number of child nodes of node v_i, then

$$D'_{i,j} = \begin{cases} (0,0) & \text{if } R = 0 \\ D_{i_1,j} & \text{if } R = 1 \\ D_{i_1,j} \oplus \cdots \oplus D_{i_R,j} & \text{if } R \geq 1 \end{cases} \tag{1.3}$$

   Then, for each k in $E_{i,k}$,

$$D_{i,j} = D'_{i,j-k} \oplus E_{i,k} \tag{1.4}$$

   //In equation (5.1), $D_{i_1,j} \oplus D_{i_2,j}$ is computed as follows. Let G' be the union of all nodes in the graphs rooted at nodes v_{i_1} and v_{i_2}. Travel all the graphs rooted at nodes v_{i_1} and v_{i_2}. For each node a in G', we add the cost of a and multiply the probability of a to $D'_{i,j}$ only once, because each node can only have one assignment and there is no assignment conflict.//

4. For each possible fixed assignment, we get a $D_{N,j}$. Merge the (Probability, Cost) pairs in all the possible $D_{N,j}$ together, and sort them in

ascending sequence according to probability. Then use the Lemma 1.3.2 to remove redundant pairs. The final $D_{N,j}$ we get is the table in which each entry has the minimum cost with a guaranteed confidence probability under the timing constraint j.

Algorithm *DAG_Opt* gives the optimal solution when the given DFG is a DAG. In the following, we give the Theorem 1.3.3 and Theorem 1.3.4 about this. Due to the space limitation, we will not give proofs for them in this chapter.

Theorem 1.3.3 *In each possible fixed assignment, for each pair* $(P_{i,j}, C_{i,j})$ *in* $D_{i,j}$ $(1 \leq i \leq N)$ *obtained by algorithm* DAG_Opt, $C_{i,j}$ *is the minimum total cost for the graph* G^i *with confidence probability* $P_{i,j}$ *under timing constraint* j.

Theorem 1.3.4 *For each pair* $(P_{i,j}, C_{i,j})$ *in* $D_{N,j}$ $(1 \leq j \leq L)$ *obtained by algorithm* DAG_Opt, $C_{i,j}$ *is the minimum total cost for the given DAG* G *with confidence probability* $P_{i,j}$ *under timing constraint* j.

According to Theorem 1.3.3, in each possible fixed assignment, for each pair $(P_{i,j}, C_{i,j})$ in $D_{i,j}$ we obtained, $C_{i+1,j}$ is the total cost of the graph G_{i+1} with confidence probability $P_{i+1,j}$ under timing constraint j. In step 4 of the algorithm *DAG_Opt*, we try all the possible fixed assignments, combine them together into a new row $D_{N,j}$ in dynamic table, and remove redundant pairs using the Lemma 1.3.2. Hence, for each pair $(P_{i,j}, C_{i,j})$ in $D_{N,j}$ $(1 \leq j \leq L)$ obtained by algorithm *DAG_Opt*, $C_{i,j}$ is the minimum total cost for the given DAG G with confidence probability $P_{i,j}$ under timing constraint j.

In algorithm *DAG_Opt*, there are K^t loops and each loop needs $O(|V|^2 * L * M * K)$ running time. The complexity of *algorithm DAG_Opt* is $O(K^t * |V|^2 * L * M * K)$, where t is the total number of nodes with multi-parent (or multi-child) in bottom up approach (or top down approach), $|V|$ is the number of nodes, L is the given timing constraint, M is the maximum number of FU types for each node, and K is the maximum number of execution time variation for each node. Algorithm *DAG_Opt* is exponential, hence it cannot be applied to a graph with large amounts of multi-parent and multi-child nodes.

In this example of Figure 1.4, the algorithm *DAG_Heu* gives the same results as those of the algorithm *DAG_Opt*. Actually, experiments shown that although algorithm *DAG_Heu* is only near-optimal, it can give the same results as those given by the optimal algorithm in most cases.

1.3.3 Experiments

This subsection presents the experimental results of our algorithms. We conduct experiments on a set of benchmarks including voltera filter, 4-stage lattice filter, 8-stage lattice filter, differential equation solver, RLS-languerre lattice filter, and elliptic filter. Among them, the DFG for first three filters are trees and those for the others are DAGs. The basic information about the benchmarks is shown in Table 1.4. Three different FU types, R_1, R_2, and R_3, are used in the system, in which a FU with type R_1 is the quickest with the highest cost and a FU with type R_3 is the slowest with the lowest cost. The execution times, probabilities, and costs for each node are randomly assigned. For each benchmark, the first timing constraint we use is the minimum execution time. The experiments are performed on a Dell PC with a P4 2.1 G processor and 512 MB memory running Red Hat Linux 7.3.

Benchmarks	**DFG**	**# of nodes**	**# of multi-parent**	**# of multi-child**
voltera	Tree	27		
4-lat-iir	Tree	26		
8-lat-iir	Tree	42		
Diff. Eq.	DAG	11	3	1
RLS-lagu.	DAG	19	6	3
elliptic	DAG	34	8	5

TABLE 1.4: The basic information for the benchmarks.

Voltera Filter							
TC	0.7		0.8		0.9		1.0
	cost	%	cost	%	cost	%	cost
62	7896		×		×		×
80	7166		7169		7847		×
100	5366	31.5	5369	31.4	6047	22.8	**7827**
125	5347	31.7	5352	31.6	5843	25.3	**7820**
150	4032	43.8	4066	43.6	4747	32.8	**7169**
175	1604	66.2	2247	52.7	2947	37.9	**4747**
200	1587	66.3	1618	65.6	2318	50.7	**4704**
225	1580	46.4	1593	45.9	1647	44.1	**2947**
250	1580	31.9	1582	31.8	1604	30.8	**2318**
273	1580	4.1	1580	4.1	1580	4.1	**1647**
274	1580		1580		1580		**1580**
Ave. Redu.(%)		40.2		38.3		31.0	

TABLE 1.5: The minimum total costs with computed confidence probabilities under various timing constraints for voltera filter.

The experiments on voltera filter, 4-stage lattice filter, and 8-stage lattice filter are finished in less than one second, in which we compare our algorithms with the *hard HA algorithms.* The experimental results for voltera filter are shown in Table 1.5. In each table, column "TC" represents the given timing constraint. The minimum total costs obtained from different algorithms, *DAG_Opt* and the optimal *hard HA algorithms* [183], are presented in each entry. Columns "1.0", "0.9", "0.8", and "0.7" represent that the confidence probability is 1.0, 0.9, 0.8, and 0.7, respectively. Algorithm *DAG_Opt* covers all the probability columns, while the optimal *hard HA algorithm* [183] only includes the column "1.0", which is in boldface. For example, in Table 1.5, under the timing constraint 100, the entry under "1.0" is 7827, which is the minimum total cost for the hard HA problem. The entry under "0.9" is 6047, which means we can achieve minimum total cost 6047 with confidence probability 0.9 under timing constraints. From the information provided in the structure of the link list in each entry of the dynamic table, we are able to trace how to get the satisfied assignment.

Column "%" shows the percentage of reduction on the total cost, comparing the results of algorithm with those obtained by the optimal *hard HA algorithms* [183]. The average percentage reduction is shown in the last row "Ave. Redu. (%)" of Table 1.5. The entry with "×" means no solution available. In Table 1.5, under timing constraint 80, the optimal *hard HA algorithms* cannot find a solution. However, we can find solution 7847 with probability 0.9 that guarantees the total execution time of the DFG is less than or equal to the timing constraint 80.

RLS-Laguerre filter							
TC	0.7		0.8		0.9		1.0
	cost	%	cost	%	cost	%	cost
49	7803		×		×		×
60	7790		7791		7793		×
70	7082		7087		7787		×
80	5403	30.6	5991	23.0	5993	23.0	**7780**
100	3969	48.9	4669	39.9	5380	30.8	**7769**
125	2165	59.8	2269	58.0	4664	13.6	**5390**
150	1564	66.4	2264	49.3	2864	38.6	**4667**
175	1564	66.5	1564	66.5	2264	51.5	**4664**
205	1564	30.9	1564	30.9	1564	30.9	**2264**
206	1564		1564		1564		**1564**
Ave. Redu.(%)		50.5		44.6		31.4	

TABLE 1.6: The minimum total costs with computed confidence probabilities under various timing constraints for rls-laguerre filter.

We also conduct experiments on RLS-Laguerre filter, elliptic filter, and differential equation solver, which are DAGs. RLS-Laguerre filter has 3 multi-child nodes and 6 multi-parent nodes. Using top-down approach, we implemented all $3^3 = 27$ possibilities. Elliptic filter has 5 multi-child nodes and 8 multi-parent nodes. There are total $3^5 = 243$ possibilities by top-down approach. Table 1.6 shows the experimental results for RLS-Laguerre filter. Column "%" shows the percentage of reduction on system cost, comparing the results for soft real-time with those for hard real-time. The average percentage reduction is shown in the last row "Ave. Redu. (%)" of all these two tables. The entry with "×" means no solution available. Under timing constraint 70 in Table 1.6, there is no solution for hard real-time. However, we can find solution 7787 with probability 0.9 that guarantees the total execution time of the DFG is less than or equal to the timing constraint 70.

The experimental results show that our algorithms can greatly reduce the total cost while having a guaranteed confidence probability satisfying timing constraints. On average, algorithm *DAG_Opt* gives a cost reduction of 33.5%

with confidence probability 0.9 under timing constraints, and a cost reduction of 43.0% and 46.6% with 0.8 and 0.7 confidence probabilities satisfying timing constraints, respectively. The experiments using *DAG_Heu* on these benchmarks are finished within several seconds and the experiments using *DAG_Opt* on these benchmarks are finished within several minutes. Algorithm *DAG_Heu* gives good results in most cases.

The advantages of our algorithms over the *hard HA algorithms* are summarized as follows. First, our algorithms are efficient and provide an overview of all possible variations of minimum costs compared with the the worst-case scenario generated by the *hard HA algorithms*. Second, it is possible to greatly reduce the system total cost while having a very high confidence probability under different timing constraints. Finally, our algorithms are very quick and practical.

1.4 Conclusion

In this chapter, we introduce some basic concepts in embedded systems designs. We illustrate the significance of the embedded systems design due to the characters of embedded systems, compared to the general purpose computing system. We also present the challengers in embedded systems designs in this chapter. For a practical challenge, the optimization in fixed time heterogenous systems, we provide two DAG-based algorithms to solve the problem.

1.5 Glossary

Adaptivity: Embedded systems are usually worked under harsh environment with: a) heat, vibration, and shock; b) power fluctuations, RF interference, and lightning; and c) water, corrosion, and physical abuse. The ability to deal with this is critical to the applications.

Architectural-Level Synthesis: Architectural-level synthesis is an automated design process that interprets an algorithmic description of a desired behavior and creates hardware that implements that behavior.

Embedded System: An embedded system is a computer system designed to perform one or a few dedicated functions often with real-time computing constraints.

Heterogeneous Computing Systems: Heterogeneous computing systems refer to electronic systems that use a variety of different types of computational units.

Performance: The execution time of the systems. The two main measures of performance are: latency and throughput, deemed adaptable when there is agreement among suppliers, owners, and customers that the process will meet requirements throughout the strategic period.

Safety: This means embedded systems must function correctly and must not function incorrectly.

Time-to-Market Design Metric: The time-to-market constraint has become especially demanding in recent years. Introducing an embedded system to the marketplace early can make a big difference in the system's profitability.

Chapter 2

Optimization for Real-Time Embedded Systems

2.1 Introduction

Heterogeneous embedded systems are driving an information revolution with their pervasion in our everyday lives. With the advance of the technology, embedded systems with multiple cores or VLIW-like architectures, such as multicore network processors, TI's TMS320C6x, Philip's TriMedia, IBM/SONY/TOSHIBA's CELL processor, Intel's newest 80 core processor, etc., become necessary to achieve the required high performance for the applications with growing complexity. While these parallel architectures can be exploited to increase parallelism and improve time performance, power optimization techniques are needed to minimize their power consumption. To optimize one objective is what most research has been doing, but a realistic embedded systems design requires efficient algorithms and tools that can handle multiple objectives simultaneously.

The design of heterogeneous embedded systems poses a host of technical challenges different from those faced by general-purpose computers because they have much larger design space and are more complicated to optimize in terms of cost, timing, and power. For example, the CELL processor is a heterogeneous, multicore chip for computation-intensive broadband media applications [81]. To optimize, a realistic embedded systems design requires efficient algorithms and tools that can handle multiple objectives simultaneously. In this chapter, we will design new techniques that can consider timing and power optimization collectively for parallel and heterogeneous embedded systems.

Design space exploration is considered as one of the most challenging problems in synthesis. This problem is particularly important and challenging for parallel and heterogeneous systems. For heterogeneous parallel embedded systems, there are more and more *functional units* (FUs) available to be selected for using. A real application with heterogeneous components, such as *digital signal processing* (DSP) architecture, usually consists of many tasks to be finished. Therefore, an important problem arises: how to assign a proper FU type to each operation in such a way that the requirements can be met and the total cost can be minimized. It is not practical to solve this problem by trying all combinations since the run time will increase exponentially with the length of the input. For example, if an application has 100 operations and there are 10 different types FUs available, it needs 10^{100} steps to try all combinations. Hence, a more efficient algorithm needs to be developed.

Furthermore, in many real applications, some tasks may not have fixed execution times. This make the scenario even more complicated. For this soft real-time scenario, we model each varied execution time as a probabilistic random variable and define it as *heterogeneous assignment with probability* (HAP) problem. The solution of the HAP problem assigns a proper FU type to each task such that the total energy is minimized while the timing constraint is satisfied with a guaranteed confidence probability. We attacked this important optimization problem and get several amazing results for the first step. A broad application can be applied to these results and it is so important that we expect to further explore it and achieve more fruitful results with loop scheduling and parallel computing techniques.

Embedded and hybrid systems (EHS) is a driving force of progress and innovation in modern era as manifested in the broad range of applications in such areas as military, science, medical, transportation, education, etc. The nation's critical infrastructures depend on embedded sensing and hybrid control systems. Heterogeneity is one of the most important features of EHS, which comprise different components, or *functional units* (FUs). Real-time and source constraint is another important feature of EHS. This project will focus on allocation and scheduling of resources to improve the performance of embedded systems in time and power aspects.

2.2 Related Work

2.2.1 Heterogeneous Embedded Systems

Embedded and hybrid systems (EHS) become more and more complicated. In many systems, such as heterogeneous parallel DSP systems [88, 19], the same type of operations can be processed by heterogeneous FUs with different costs. The cost of embedded systems may relate to power, reliability, etc. Therefore, an important problem arises: how to assign a proper FU type to each operation of a DSP application such that the requirements can be met and the total cost can be minimized while satisfying timing constraints with a guaranteed confidence probability [183].

Furthermore, we observe that some tasks may not have fixed execution time. Such tasks usually contain conditional instructions and/or operations that could have different execution times for different inputs [207, 247, 92, 91, 90]. Although many static assignment techniques can thoroughly check for the best assignment for dependent tasks, existing methods are not able to deal with such uncertainty. Therefore, either worst-case or average-case computation times for these tasks are usually assumed. Such assumptions, however, may not be applicable for real-time systems and may result in an ineffective task assignment.

We call the first scenario (with fixed execution time) as hard HA (*Heterogeneous Assignment*) problem and the second scenario (with non-fixed execution time) as HAP (*Heterogeneous Assignment with Probability*) problem [164, 161]. Using probabilistic approach, we can obtain solutions that not only can be used for hard real-time systems, but also provide more choices of smaller total costs while satisfying timing constraints with guaranteed confidence probabilities.

It is a critical issue to do system-level synthesis for special purpose architectures of real-time embedded systems to satisfy the time and cost requirements [73, 116, 229, 155, 218, 97, 43, 98, 44]. The cost of embedded systems may relate to power, reliability, etc. [64, 196, 185, 85].

There have been a lot of research efforts on allocating applications in heterogeneous distributed systems [23, 88, 19, 174, 27, 22, 24, 21]. Incorporating reliability cost into heterogeneous distributed systems, the reliability driven assignment problem has been studied in [64, 196, 185, 85]. In this work, allocation is performed based on a fixed architecture. However, when performing assignment and scheduling in architecture synthesis, no fixed architectures are available. Most previous work on the synthesis of special purpose architectures for real-time DSP applications focuses on the architectures that only use homogeneous FUs, that is, same type of operations will be processed by same type of FUs [155, 97, 132, 130, 149, 93, 77, 44, 40, 120]. In [183], we have proposed two optimal algorithms for the hard HA problem when the given input is a tree or simple path, and three heuristic algorithms for the general hard HA problem. But we do not consider the varied execution time situation. Also, for the general problem, the solutions are not optimal.

Probabilistic retiming (PR) had been proposed by Tongsima et al. in [207, 153]. For a system without resource constraints, PR can be applied to optimize the input graph, i.e., reduce the length of the longest path of the graph such that the probability of the longest path computation time being less than or equal to the given timing constraint, L, is greater than or equal to a given confidence probability P. Since the execution times of the nodes can be either fixed or varied, a probability model is employed to represent the execution time of the tasks. But PR does not model the hard HA problem which focuses on how to obtain the best assignment from different FU types.

In paper [98], the authors proposed the ILP model to obtain an optimal solution for the heterogeneous assignment problem. But it is a NP-hard problem to solve the ILP model. Therefore, the ILP model may take a very long time to get results even when a given *Data Flow Graph* (DFG) is not very big. In this project, the algorithm solving the hard HA problem in [98] is called *hard HA ILP algorithm* [98] in general.

We present system-level synthesis algorithms which operate in probabilistic environments to solve the *heterogeneous assignment with probability* (HAP) problem. In the HAP problem, we model the execution time of a task as a random variable [241]. For heterogeneous systems, each FU type has a different cost, representing hardware cost, size, reliability, etc. Faster one has higher cost while slower one has lower cost. In papers [160, 164, 161, 165], we show how to assign a proper FU type to each node of a *Probabilistic Data Flow Graph* (PDFG) such that the total cost is minimized while satisfying the timing constraint with a guaranteed confidence probability. In other words, we can guarantee that the total execution time of the PDFG is less than or equal to the timing constraint with a probability greater than or equal to a constant probability P.

It is known that the hard HA problem is NP-complete [183, 76]. Since the HAP problem is NP harder than the hard HA problem, the HAP problem is also NP-complete. We will use the proposed pseudo polynomial time algorithms to optimally solve the HAP problem when the given PDFG is a tree or a simple path, and two other algorithms are proposed to solve the general problem, which is a DAG (*Directed Acyclic Graph*) [160, 164, 161, 165]. In this chapter, we will continue to attack this important problem by using loop scheduling and parallel computing (multiprocessor) techniques. Time and power optimization is our first target.

2.2.2 Time and Power Optimization

In recent years, system-level synthesis and optimization have drawn immense attentions due to the demand for rapid time-to-market and the performance requirement of special purpose systems for applications such as DSP, image processing, radar, telecommunications and medical systems. Many applications can be modeled as various classes of graphs such as *data-flow graphs* (DFG). Then, the code transformation, scheduling and optimization problems can be regarded as graph theoretical ones. In order to avoid an ad-hoc design, it is extremely important to study the fundamentals of the problems regarding the properties, complexities and the designs of optimal algorithms.

It's important to accomplish fundamental works in synthesis and optimization areas. Leiserson and Saxe [115] have developed the Retiming algorithm which gave a very significant contribution in graph transformations and timing optimization. Sha [41, 44, 43, 239] developed loop scheduling in parallel systems and designed the polynomial-time algorithms which produce the rate-optimal schedule for any given data-flow graph assuming enough resources are available.

Embedded systems become increasingly heterogeneous, which pose great challenges different from those faced by general-purpose computers. Embedded systems are more application specific and more constrained in terms of power, timing, and other resources. Energy-saving [98, 25, 190, 50, 124, 240, 156, 16, 119, 236, 99, 243, 140, 170, 135, 96, 97, 219] is a critical issue and performance metric in embedded systems design due to wide use of portable devices, especially those powered by batteries [73, 116, 229, 155].

The systems become more and more complicated and some tasks may not have fixed execution times. Such tasks usually contain conditional instructions and/or operations that could have different execution times for different inputs [207, 247, 92, 91, 90]. It is possible to obtain the execution time distribution for each task by sampling and knowing detailed timing information about the system or by profiling the target hardware [205]. Also some multimedia applications, such as image, audio, and video data streams, often tolerate occasional deadline misses without being noticed by human visual and auditory systems. For example, in packet audio applications, loss rates between 1% – 10% can be tolerated [30].

Prior design space exploration methods for hardware/software codesign of DSP systems [218, 97, 98, 183] guarantee no deadline missing by considering worst-case execution time of each task. Many design methods have been developed based on worst-case execution time to meet the timing constraints without any deadline misses. These methods are pessimistic and are suitable for developing systems in a hard real-time environment, where any deadline miss will be catastrophic. However, there are also many soft real-time systems, such as heterogeneous systems, which can tolerate occasional violations of timing constraints. The above pessimistic design methods can't take advantage of this feature and will often lead to over-designed systems that deliver higher performance than necessary at the cost of expensive hardware, higher energy consumption, and other system resources.

There are several papers on the probabilistic timing performance estimation for soft real-time systems design [161, 247, 100, 205, 92, 91, 90, 207]. The general assumption is that each task's execution time can be described by a discrete probability density function that can be obtained by applying path analysis and system utilization analysis techniques. Hu et al. [247] propose a state-based probability metric to evaluate the overall probabilistic timing performance of the entire task set. However, their evaluation method becomes very time consuming when the task has many different execution time variations. Hua et al. [92, 91] propose the concept of *probabilistic design* where they design the system to meet the timing constraints of periodic applications statistically. But their algorithm is not optimal and only suitable to uniprocessor executing tasks according to a fixed order, that is, a simple path.

Dynamic voltage scaling (DVS) is one of the most effective techniques to reduce energy consumption [53, 187, 244, 180, 52]. In many microprocessor systems, the supply voltage can be changed by mode-set instructions according to the workload at run-time. With the trend of multiple cores being widely used in DSP systems, it is important to study DVS techniques for multiprocessor DSP systems.

Many researches have been done on DVS for real-time applications in recent years [187, 244, 180]. Zhang et al. [244] proposed an ILP (Integer Linear Programming) model to solve DVS on multiple processor systems. Shin et al. [187] proposed a DVS technique for real-time applications based on static timing analysis. However, in the above work, applications are modeled as DAG (Directed Acyclic Graph), and loop optimization is not considered. Saputra et al. [180] considered loop optimization with DVS. However, in their work, the whole loop is scaled with the same voltage. Our technique can choose the best voltage level assignment for each task node.

We focus on minimizing expected energy consumption with guaranteed probability satisfying timing constraints via DVS for real-time multiprocessor embedded systems. We use probabilistic design space exploration and DVS to avoid over-designing systems. Our work is related to the work in [92, 91]. In [92, 91], Hua et al. proposed a heuristic algorithm for uniprocessor and the PDFG (Probabilistic Data Flow Graph) is a simple path. We call the offline part of it as *HUA* algorithm for convenience. In [160], we also applied the greedy method of *HUA* algorithm to multiprocessor and called the new algorithm produced as *Heu*. We compared our algorithms with Hua's algorithm for uniprocessor and *Heu* algorithm. We proposed two novel optimal algorithms, one for uniprocessor and one for multiprocessor embedded systems, to minimize the expected value of total energy consumption while satisfying timing constraints with guaranteed probabilities for real-time applications [160].

In [184, 182, 183, 53, 98], our lab developed the efficient assignment and scheduling technique to optimize the energy/reliability of heterogeneous DSP systems at architecture level. The experimental results show that our recommended algorithm can achieve 27.9% improvement on average in energy/reliability compared with the previous work. We have developed efficient algorithms for heterogeneous embedded systems with probabilistic execution time model [160, 164, 161, 165]. In this chapter, we will further optimize the heterogeneous embedded systems with loop scheduling and multicore techniques.

2.3 Graph Models and Timing Optimization

2.3.1 Basic Graph Models and Techniques

Many applications can be modeled with data-flow graphs (DFG) or other sequencing graphs. Thus many important phases of synthesis such as graph transformation, scheduling and optimization problems can be regarded as graph-theoretical ones.

Definition 2.1 *A data-flow graph DFG* $G = \langle V, E, d, t\rangle$ *is a node-weighted edge-weighted directed graph where* V *is a set of nodes,* E *is a set of edges,* d *is a function from* E *to* $\mathbb{N}$ *representing a number of delays on each edge, and* $t(v)$ *represents computation time of each node.*

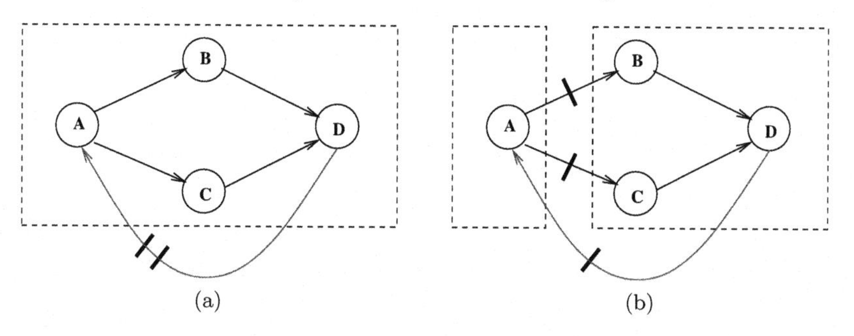

FIGURE 2.1: (a) A data-flow graph. (b) Retimed graph $r(A) = 1$.

As an example, Figure 2.1(a) shows a simple data-flow graph where $V = \{A, B, C, D\}$. We use a shorthand $u \to v$ to represent an edge from node u to node v. The edge set in this graph is composed of five edges. Among them, edge $D \to A$ has two delays, which are denoted by bar lines on the edge, and the other edges have no delay. The computation time of each node usually is represented by a number beside a node. In this simplified example, we assume each node takes 1 time unit to compute.

An ***iteration*** is the execution of each node in V exactly once. Inter-iteration dependencies are represented by weighted edges. An iteration is associated with a static schedule. A static schedule must obey the precedence relations defined by the DFG. For any iteration j, an edge e from u to v with delay $d(e)$ conveys that the computation of node v at iteration j depends on the execution of node u at iteration $j - d(e)$. An edge with no delay represents a data dependency within the same iteration.

We use the notations $D(p)$ and $T(p)$ to represent the total computation time and total delay count of path p, respectively. The ***cycle period*** $c(G)$ of a data-flow graph G is the computation time of the longest zero-delay path; in other words $c(G) = \max\{T(p) : p \in G \text{ is a path with } D(p) = 0\}$. For example, $c(G)$ of the graph in Figure 2.1(a) is 3.

In the scheduling community, given a DFG G which may contain cycles, people developed loop scheduling algorithms to minimize the execution time of all tasks in one iteration. We have produced significant results for the optimization of loop scheduling using both retiming and unfolding.

Retiming [115] is one of the most effective graph transformation techniques for optimization. It transforms a DFG to minimize its cycle period in polynomial time by redistributing registers in a circuit (or equivalently, delays in the DFG). In the scheduling or code generation area, we have shown that a so-called *software pipelining* can be correctly modeled as a retiming. With this retiming perspective, the software pipelining can be easily analyzed from a graph-theoretical view. A *retiming* r is a function from V to $\mathbb{Z}$ that redistributes the nodes in the original DFG G, resulting in a new

DFG $G_r = \langle V, E, d_r, t \rangle$ such that each iteration still has one execution of each node in G. The delay function changes accordingly to preserve dependencies, i.e., $r(v)$ represents delay units pushed into the edges $v \to w$, and subtracted from the edges $u \to v$, where $u, v, w \in G$. Therefore, we have $d_r(e) = d(e) + r(u) - r(v)$ for every edge $u \to v$ and $d_r(\ell) = d(\ell)$ for every cycle $\ell \in G$. As an example, Figure 2.1(b) portrays the retimed graph G_r of the DFG G in Figure 2.1(a) where $r(A) = 1$ and $r(B) = r(C) = r(D) = 0$. A cycle period $c(G_r)$ now becomes 2 instead of 3 in the DFG in Figure 2.1(a).

Unfolding [44, 149] is another effective technique for improving the average cycle period of a static schedule. The original DFG G is unfolded f times, so the unfolded graph G_f consists of f copies of the original node set. Thus, a schedule with unfolding factor f contains f iterations of the original DFG. The instruction-level parallelism between these iterations helps to improve the iteration period $P = c(G_f)/f$.

For any DFG G, the average computation time of an iteration is called the ***iteration period*** of the graph. If a DFG contains a loop, then the iteration period is bounded from below by the iteration bound [176] of the graph which is defined as follows:

Definition 2.2 *The* ***iteration bound*** *of a data-flow graph G, denoted by $B(G)$, is the maximum time to delay ratio of all cycles in G. This can be represented by the equation $B(G) = \max_{\forall l} \frac{T(l)}{D(l)}$, where $T(l)$ and $D(l)$ are the summation of all computation times and the summation of delays, respectively, in a cycle l.*

Consider the graph G in Figure 2.1(b). The iteration bound of this graph is therefore $1\frac{1}{2}$. It is clear that unfolding will increase code size significantly. We want to find the minimum f with retiming r so the iteration period of the resultant loop schedule is optimal. But it is very likely that such an optimal f is too large for the program to fit into a small-size on-chip memory; then we need to explore what will be good f, r, schedule and code size.

2.3.2 Timing Optimization

The relations between retiming, unfolding and scheduling were explored by Sha et al. [44]. Two types of scheduling problems are studied: scheduling with resource constraints and scheduling without resource constraints. For many hardware platforms such as new TI DSP VLIW processors allowing at most eight operations to be executed in parallel, the resource constraint is necessary to be considered in scheduling. We have developed a polynomial-time scheduling algorithm with resource constraints, called ***rotation scheduling***, which uses retiming on data-flow graph as a basic operation, and iteratively produces a close-to-optimal schedule in polynomial time. Its high performance has been proved by many of our experiments. *Modulo scheduling* is probably the most popular technique which is currently employed in many platforms. It is interesting to note by our recent experiments that the rotation scheduling algorithm consistently outperforms modulo scheduling for DSP applications. However, rotation scheduling algorithm only optimizes the schedule length. A lot of improvement and various extensions must be made to design new scheduling algorithms for multiple optimization objectives [250, 181].

It is known that retiming and unfolding techniques can be used jointly to produce schedules with good average iteration period [44]. Let r be the retiming value, f the unfolding factor, S the number of points in design space. The design space considering various retiming values and unfolding factors is $r \times f \times S$. The challenging problem is how to find ***feasible*** design points in this large design space within a short time. In order to reduce such a space, it is important to understand some fundamental relations: what are the feasible unfolding factors that can achieve the iteration period requirement by combining with retiming? Then based on this understanding, we can produce **the minimum set of unfolding factors** for achieving performance requirement. Intuitively speaking, the obtained interrelation *reduces the points to be selected from a higher-dimensional volume to a lower-dimensional plane.*

We derived Theorem 2.3.1 and Theorem 2.3.2 to find the minimum possible unfolding factor that can achieve the iteration period constraint via retiming. We can use these two theorems to easily check (without real scheduling) if an unfolding factor is possible to satisfy timing requirement or not.

Theorem 2.3.1 *Let $G = \langle V, E, d, t \rangle$ be a given data flow graph, $f \in Z^+$ an unfolding factor, and $c \in R$ a cycle period, there exists a legal static schedule (with retiming) iff $c/f \geq B(G)$ and $c \geq max_v t(v)$, $\forall v \in V$. Thus, given unfolding factor f, the minimum cycle period $c_{min}(G_f) = max(max_v t(v), \lceil f \cdot B(G) \rceil)$.*

Consider a simple DFG composed of three nodes, node A, B and C, with computation time 1, 5, 1 time units, respectively, and three edges, $A \rightarrow B$ and $B \rightarrow C$ with no delay, and edge $C \rightarrow A$ with four delays. Then, the cycle period of the DFG is $c(G) = 1 + 5 + 1 = 7$ time units, and the iteration bound is $B(G) = 7/4$. If the unfolding factor $f = 2$, we can directly find the best feasible cycle period combined with retiming by Theorem 2.3.1, that is, $c_{min}(G_f) = max(5, 4) = 5$. Then it proves that it is impossible to find a schedule with cycle period < 5 with $f = 2$. This is a very clear and strong result.

Theorem 2.3.2 *Let $G = \langle V, E, d, t \rangle$ be a data flow graph, f an unfolding factor, P a given iteration period constraint. The following statements are equivalent:*

1. *There exists a legal static schedule of unfolded graph G_f with iteration period less than or equal to P.*

2. *$B(G) \leq c_{min}(G_f)/f \leq P$.*

As to the previous example, assume that we want to achieve an average iteration period $P = 7/3$, and we would like to know what unfolding factor is possible for achieving this requirement. For instance, how about unfolding factor $f = 2$? Since the cycle period of unfolded graph $c_{min}(G_f) = 5$ (from Theorem 2.3.1), and we find that $c_{min}(G_f) > P \cdot f = 4\frac{2}{3}$, so the iteration period constraint cannot be achieved with unfolding factor $f = 2$. By using Theorem 2.3.2, we can immediately eliminate many infeasible ones *without performing real scheduling*, and thus significantly reduce the design space.

Since retiming and unfolding greatly expand the code size [250, 252], it's possible that the generated code is too large to be fit into the on-chip memory. Thus, the relationship between schedule length and code size must be studied. The relationship between code size and performance can be formulated by mathematical formula using retiming functions. It provides an efficient technique for reducing the code size of any retimed graph.

Benchmarks (Algo.)	Search Points	Ratio	Final Solutions				
			uf	#A	#M	IP	CS
Biquad(std.)	228		4	4	16	3/2	80
Biquad(ours)	**4**	**1.5%**	**3**	**3**	**10**	**5/3**	**28**
DEQ(std.)	486		5	10	18	8/5	110
DEQ(ours)	**6**	**1.2%**	**3**	**5**	**15**	**5/3**	**37**
Allpole(std.)	510		5	10	10	18/5	150
Allpole(ours)	**6**	**1.2%**	**3**	**10**	**3**	**11/3**	**51**
Elliptic(std.)	694		F	F	F	F	F
Elliptic(ours)	**2**	**0.3%**	**2**	**6**	**9**	**5**	**76**
4-Stage(std.)	909		F	F	F	F	F
4-Stage(ours)	**55**	**6%**	**4**	**7**	**33**	**7/4**	**112**

TABLE 2.1: Comparing our design space exploration algorithm with standard method on DSP benchmarks.

Table 2.1 illustrates the quality and performance differences by comparing our design exploration algorithm against standard search method. Our algorithm utilizes unfolding and extended retiming, a new retiming algorithm developed by Sha et al. [143, 142], which gives a better result than traditional retiming. Column "Search Point" represents the number of design points that have been searched. In "Final Solution" column, "uf" denotes the unfolding factor for achieving the iteration period ("Iter. Period"). Column "#A" represents the number of adders. Column "#M" represents the number of multipliers. The minimum configuration (the required adders and multipliers) computed by algorithms are shown in columns "#A" and "#M" in the table. Column "IP" represents iteration period and column "CS" represents code size.

The table entry "F" indicates a failure of search for feasible solution. As to the **effectiveness of design space minimization**, it shows that the search space size using our method is only 2% of the search space using the standard method on average. Also, the **quality** of the solutions found by our algorithm are better than that of standard methods because ours requires less number of adders, multipliers and smaller code size.

These experimental results are very promising. Not only the design space is greatly reduced, but the quality of design solutions are much better than those generated by standard methods. This good result confirms the importance of fundamental research on properties and relations between design parameters for synthesis problem.

Probabilistic Data-Flow Graph

The graph model study will also include new representations of the traditional data-flow graph, which is extensively used in embedded system synthesis. For example, the probabilistic data-flow graph (PDFG) will be studied to accurately convey the system behavior of a complex system. In many practical applications such as interface systems, fuzzy systems, and artificial intelligence systems, the required tasks normally have uncertain computation times. Existing methods are not able to deal with such uncertainty; therefore, either worst-case or average-case computation time for this type of task is usually assumed. Such assumptions, however, may not be applicable for the real operating situation and may result in an inefficient schedule. The computation time and communication time become random variables. The objectives of using the probabilistic approach is to obtain the optimized graph which has its cycle period close to optimal with a high degree of confidence. According to our experiment, the use of the new probabilistic retiming results in the improvement of schedule length up to **40%** over those from traditional retiming considering worst-case scenario and up to **30%** when considering average-case scenario. Therefore, the advantage is clear to design optimization algorithms specifically considering the probabilistic models. To design a system that can be proved to operate well with **a high probability** under all possible values of the event variables is not easy. In this chapter, we will design efficient design and optimization algorithms to build robust embedded systems on probabilistic models.

2.3.3 Time and Power Optimizations

2.3.3.1 Loop Scheduling and Parallel Computing

Loops are usually the most time-consuming and power-consuming parts of applications. Therefore, we will develop effective loop optimization techniques to reduce schedule length, memory access, dynamic power, static power and energy with dynamic voltage scaling. We have conducted several studies for low power optimization [160, 164, 161, 165]. Based on our previous work, we will develop new techniques for dynamic power optimization, static power optimization, dynamic voltage scaling and memory access optimization.

We use probabilistic approach and loop scheduling to avoid over-design systems. In this project, we will develop a novel optimization algorithm to minimize expected value of total energy consumption while satisfying timing constraints with guaranteed probabilities for real-time applications. We will define the VASP *(voltage assignment and scheduling with probability)* in this subsection and solve it through this project.

We design new rotation scheduling algorithms for real-time applications that produce schedules consuming minimal energy. In our algorithms, we use rotation scheduling [41, 44] to get schedules for loop applications. The schedule length will be reduced after rotation. Then, we use *dynamic voltage scaling* (DVS) to assign voltages to computations individually in order to decrease the voltages of processors as much as possible within the timing constraint.

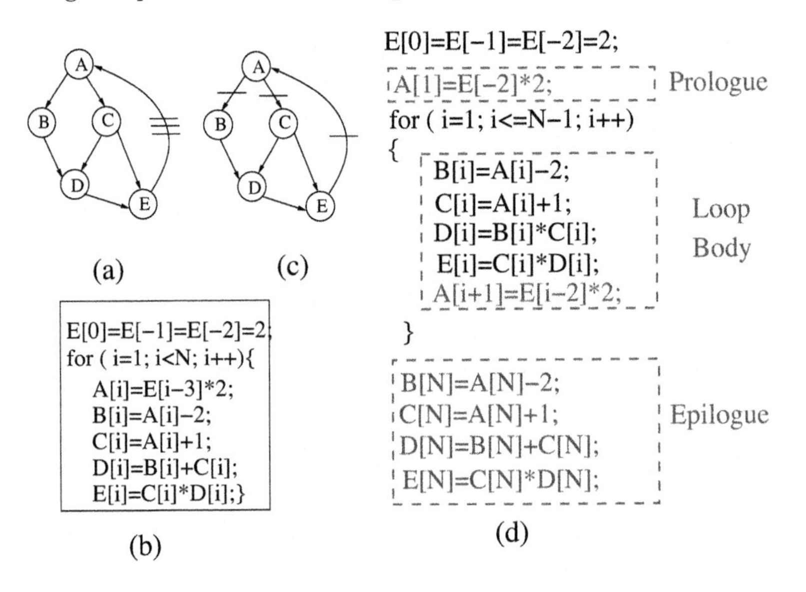

FIGURE 2.2: (a) The original PDFG. (b) The code of loop application corresponding to the original PDFG. (c) The rotated PDFG. (d) The equivalent loop after regrouping loop body.

Static Schedules: From the PDFG of an application, we can obtain a static schedule. A static schedule [43] of a cyclic PDFG is a repeated pattern of an execution of the corresponding loop. In our works, a schedule implies both control step assignment and allocation. A static schedule must obey the dependency relations of the *Directed Acyclic Graph* (DAG) portion of the PDFG. The DAG is obtained by removing all edges with delays in the PDFG.

Retiming: Retiming [115] is an optimal scheduling technique for cyclic PDFGs considering inter-iteration dependencies. It can be used to optimize the cycle period of a cyclic PDFG by evenly distributing the delays. Retiming generates the optimal schedule for a cyclic PDFG when there is no resource constraint. Given a cyclic PDFG G=⟨U, ED, d, T, V⟩, retiming r of G is a function from U to integers. For a node $u \in U$, the value of r(u) is the number of delays drawn from each of incoming edges of node u and pushed to all of the outgoing edges. Let G_r=⟨U, ED, d_r, T, V⟩ denote the retimed graph of G with retiming r, then $d_r(ed) = d(ed) + r(u) - r(v)$ for every edge $ed(u \rightarrow v) \in$ ED.

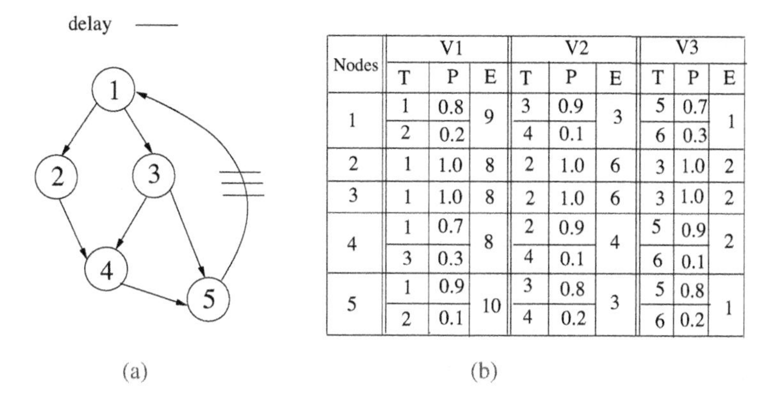

Nodes	V1			V2			V3		
	T	P	E	T	P	E	T	P	E
1	1	0.8	9	3	0.9	3	5	0.7	1
	2	0.2		4	0.1		6	0.3	
2	1	1.0	8	2	1.0	6	3	1.0	2
3	1	1.0	8	2	1.0	6	3	1.0	2
4	1	0.7	8	2	0.9	4	5	0.9	2
	3	0.3		4	0.1		6	0.1	
5	1	0.9	10	3	0.8	3	5	0.8	1
	2	0.1		4	0.2		6	0.2	

FIGURE 2.3: (a) A PDFG. (b) The times, probabilities, and energy consumptions of its nodes under different voltage levels.

Rotation Scheduling: Rotation scheduling [41] is a scheduling technique used to optimize a loop schedule with resource constraints. It transforms a schedule to a more compact one iteratively in a PDFG. In most cases, the minimal schedule length can be obtained in polynomial time by rotation scheduling. Figure 2.2 shows an example to explain how to obtain a new schedule via rotation scheduling. We use the schedule generated by list scheduling in Figure 2.3(a) as an initial schedule. In Figure 2.2(a) we change the node label 1, 2, 3, 4, 5 to A, B, C, D, E. Figure 2.2(b) shows the corresponding code. We get a set of nodes at the first row of the schedule, in this case, it is {1}, and we rotate node 1 down. The rotated graph is shown in Figure 2.2(c). The equivalent loop body after rotation is shown in Figure 2.2(d). The code size is increased by introducing the prologue and epilogue after the rotation is performed, which can be solved by the technique proposed in [251].

Parallel Computing: A parallel computing system is a computer with more than one processor for parallel processing. In the past, each processor of a multiprocessing system always came in its own processor packaging, but recently introduced multicore processors contain multiple logical processors in a single package. There are many different kinds of parallel computers. They are distinguished by the kind of interconnection between processors (known as "processing elements" or PEs) and memory [116, 38, 139, 125]. Researchers have explored several design avenues in both academia and industry. Examples include MIT's Raw multiprocessor, the University of Texas's Trips multipro-

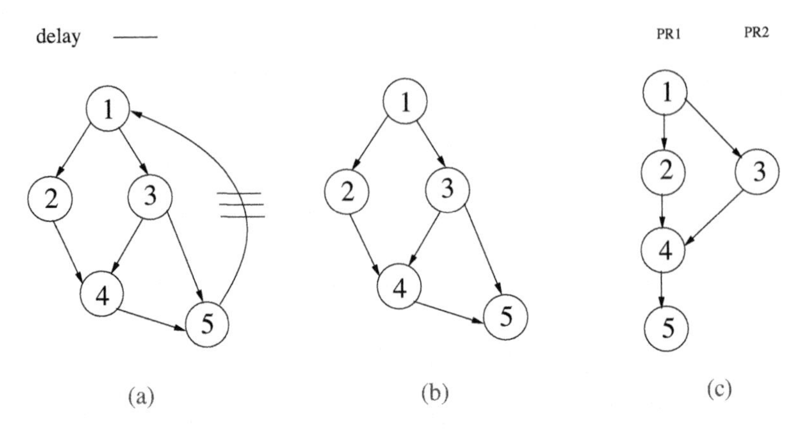

FIGURE 2.4: (a) Original PDFG. (b) The static schedule (DAG). (c) The schedule graph using list scheduling.

cessor, AMD's Opteron, IBM's Power5, Sun's Niagara, and Intel's Montecito, among many others [107]. In this project, we will exploit parallelism of multi-core embedded systems.

Definitions: Define the *VASP (voltage assignment and scheduling with probability)* problem as follows: Given M different voltage levels: V_1,V_2,$\cdots$,V_M, a PDFG $G = \langle U, ED, T, V \rangle$ with $T_{V_j}(u)$, $P_{V_j}(u)$, and $E_{V_j}(u)$ for each node $u \in U$ executed on each voltage V_j, a timing constraint L and a confidence probability P, find the voltage for each node in assignment A that gives the *minimum expected total energy consumption E with confidence probability P under timing constraint L*. In this project, we will solve the VASP problem by loop scheduling and parallel computing techniques.

2.3.3.2 Motivational Examples

Assume an input PDFG (*Probability Data Flow Graph*) shown in Figure 2.3(a). Each node can select one of the three different voltages: V_1, V_2, and V_3. The execution times (T), corresponding probabilities (P), and expected energy consumption (E) of each node under different voltage levels are shown in Figure 2.3(b). The input PDFG has five nodes. Node 1 is a multi-child node, which has two children: 2 and 3. Node 5 is a multi-parent node, and has two parents: 3 and 4. The execution time T of each node is modeled as a random variable. For example, When choosing V_1, node 1 will be finished in 1 time unit with probability 0.8 and will be finished in 2 time units with probability 0.2.

Figure 2.4(b) shows the static schedule of (a), that is, the DAG without delay edge. Figure 2.4(c) shows the schedule graph of (b) using list scheduling. For schedule graph in Figure 2.4(b), the minimum total energy consumptions

	PR1	PR2
1	1	
2		
3	3	2
4		
5	4	
6		
7		
8		
9	5	
10		
11		

(a)

	PR1	PR2
1	3	2
2		
3	4	1
4		
5		
6		
7	5	
8		
9		
10		
11		

(b)

FIGURE 2.5: (a) The template for assignment. (b) The template for rotation.

with computed confidence probabilities under the timing constraints are shown in Table 2.2. The results are generated by our algorithm, *VAP_M*, which is a sub-algorithm of our *VASP_RS* algorithm. The entries with probability that is equal to 1 (see the entries in boldface) actually give the results to the hard real-time problem which shows the worst-case scenario. For each row of the table, the E in each (P, E) pair gives the minimum total energy consumption with confidence probability P under timing constraint j.

For example, using our algorithm, at timing constraint 11, we can get (0.90, 20) pair. The assignments are shown as "Assi_1" in Table 2.3. Assignment $A(v)$ represents the voltage selection of each node v. Hence, we find the way to achieve minimum total energy consumption 20 with probability 0.90 satisfying timing constraint 11. While using the ILP and heuristic algorithm in [183], the total energy consumption obtained is 32. The assignments are shown as "Assi_2" in Table 2.3.

Figure 2.5(a) shows the assignment template with timing constraint 11 before rotation. This template gives us the idea of the relative positions of nodes in schedule graph. Figure 2.5(b) shows the new relative positions of nodes after rotating down node 1. In Figure 2.6, (a) shows the retimed PDFG, (b) is the static schedule of (a), (c) is the graph after rotation.

Table 2.4 shows the minimum total energy consumptions with confidence probabilities under different timing constraints for the new schedule graph. At timing constraint 11, we get (0.90, 10) and (1.00, 12) pairs. The detail assignment of each node is shown in Table 2.5. For (0.90, 10) pair, node 5's type was changed to V_3, then the T was changed from 4 to 6, and energy consumption change from 3 to 1. Hence the total execution time is 11, and

T	(P , E)	(P , E)	(P , E)	(P , E)	(P , E)
4	0.50, 43				
5	0.65, 39				
6	0.65, 35	0.81, 39			
7	0.65, 27	0.73, 33	0.81, 35	0.90, 39	
8	0.81, 27	0.90, 35	**1.00, 43**		
9	0.58, 20	0.73, 21	0.81, 27	0.90, 32	**1.00, 39**
10	0.72, 20	0.81, 21	0.90, 28	**1.00, 36**	
11	0.65, 14	0.90, 20	**1.00, 32**		
12	0.81, 14	0.90, 20	**1.00, 28**		
13	0.65, 12	0.90, 14	**1.00, 20**		
14	0.81, 12	0.90, 14	**1.00, 20**		
15	0.50, 10	0.90, 12	**1.00, 14**		
16	0.72, 10	0.90, 12	**1.00, 14**		
17	0.90, 10	**1.00, 12**			
18	0.50, 8	0.90, 10	**1.00, 12**		
19	0.72, 8	**1.00, 10**			
20	0.90, 8	**1.00, 10**			
21	**1.00, 8**				

TABLE 2.2: Minimum total energy consumptions with computed confidence probabilities under various timing constraints.

		Node id	T	V Level	Prob.	Energy
Assi_1	$A(u)$	1	2	V_1	1.00	9
		2	3	V_3	1.00	2
		3	3	V_3	1.00	2
		4	2	V_2	0.90	4
		5	4	V_2	1.00	3
	Total		11		0.90	20
Assi_2	$A(u)$	1	2	V_1	1.00	9
		2	1	V_1	1.00	8
		3	1	V_1	1.00	8
		4	4	V_2	1.00	4
		5	4	V_2	1.00	3
	Total		11		1.00	32

TABLE 2.3: The assignments with timing constraint 11.

the total energy consumption is 10. So the improvement of energy saving is 50.0% while the probability is still 90%. For (1.00, 12) pair, the execution times of nodes 2, 3 were changed from 1 to 3 and node 1's was changed from 2 to 6. The total energy consumptions were changed to 12, and the total execution time is still 11. Hence compared with original 32, the improvement of total energy saving is 62.5%.

If we consider the energy consumptions of switch activity, we can get more practical results. For example, after rotation once, node 1 has changed from processor $PR1$ to $PR2$. Assume the energy consumption of this switch is 1, then the final total energy consumption is 13. The energy saving is 59.4% compared with previous scheduling and assignment.

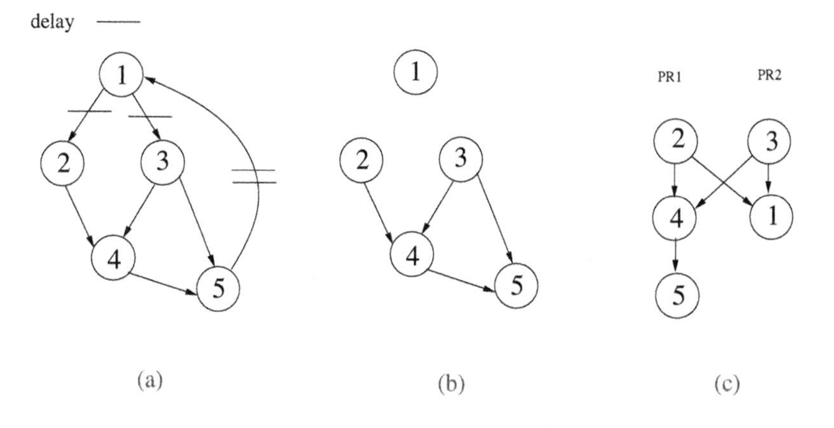

FIGURE 2.6: (a) The PDFG after retiming. (b) The static schedule. (c) The schedule after rotation.

T	(P , E)	(P , E)	(P , E)	(P , E)
3	0.63, 43			
4	0.73, 33	0.81, 39		
5	0.73, 29	0.90, 33		
6	0.73, 21	0.90, 29	**1.00, 37**	
7	0.50, 20	0.90, 24	**1.00, 31**	
8	0.50, 12	0.72, 14	0.90, 20	**1.00, 28**
9	0.90, 12	**1.00, 24**		
10	0.72, 10	0.90, 12	**1.00, 20**	
11	0.90, 10	**1.00, 12**		
12	0.90, 10	**1.00, 12**		
13	0.72, 8	**1.00, 10**		
14	0.90, 8	**1.00, 10**		
15	**1.00, 8**			

TABLE 2.4: Minimum total energy consumptions with computed confidence probabilities under various timing constraints.

2.4 Conclusion

In this chapter, we introduce some concepts about the real-time constraint and the power consumption in embedded systems designs. Furthermore, we present some advanced graph-based optimization approaches, such as retiming and unfolding. We also show the scheduling methods focused on loop architectures in parallel computing.

		Node id	T	V	Prob.	Energy
Assi_1	$A(v)$	1	6	V_3	1.00	1
		2	3	V_3	1.00	2
		3	3	V_3	1.00	2
		4	2	V_2	0.90	4
		5	6	V_3	1.00	1
	Total		11		0.90	10
Assi_2	$A(v)$	1	6	V_3	1.00	1
		2	3	V_3	1.00	2
		3	3	V_3	1.00	2
		4	4	V_2	1.00	4
		5	4	V_2	1.00	3
	Total		11		1.00	12

TABLE 2.5: The assignments with timing constraint 11.

2.5 Glossary

DAG: Directed acyclic graph, a type of graph used in computer science and mathematics.

DFG: Data flow graph.

Digital Signal Processing Architecture: A specialized microprocessor with an optimized architecture for the fast operational needs of digital signal processing.

Functional Unit: A part of a CPU that performs the operations and calculations called for by the computer program.

Heterogeneous Assignment with Probability: Given M different FU types: $R_1, R_2, \ldots, R_M$, a DFG $G = < V, E >$, where $V =< v_1, v_2, \ldots, v_N >$, $R_{R_J}(v)$, $P_{R_J}(v)$, $C_{R_J}(v)$ for each node $v \in V$ executed on each FU type j, and a timing constraint L, find an assignment for G that gives the minimum total cost C with confidence probability P under timing constraint L.

Parallel Architecture: A computer architecture in which many operations are carried out simultaneously.

Chapter 3

Multicore Embedded Systems Design

3.1 Introduction

Today's embedded systems processor contains multiple processing cores and multiple memory modules (banks), as shown in Figure 3.1. The whole embedded system may have multiple processors with local and remote memory. Each memory module contains several memory units. There is special hardware called *Special Hardware Memory Units* (SHMU) inside each processor. A shared bus is used to exchange data in the system. The on-chip memory has a tight memory size constraint and fast access velocity while the remote memory (with multi-port) is larger and slower.

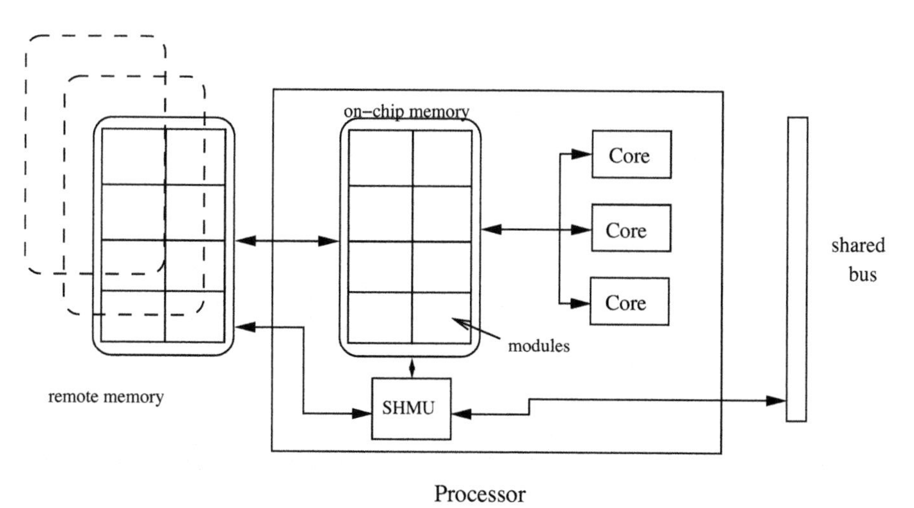

FIGURE 3.1: Architecture model.

Our architecture model is an abstract view of modern processors, such as the CELL processor. CELL processor is a heterogeneous multicore processor, with one POWER processing core (two threads) dedicated to the operating system and other control functions, and eight synergistic processing cores optimized for compute-intensive applications [86]. There is a 256k on-chip memory called local store inside each core. A total of 16 concurrent memory operations are possible at one time for each core. With this type of architecture, CELL processor gains more than 10 times the performance of a top modern CPU.

Memory latency is becoming an increasingly important performance bottleneck as the gap between processor and memory speeds continues to grow. Over the last 30 years, the processing time of a processor has increased more rapidly than the speed to access the memory. As a result of this trend, a number of techniques have been devised in order to hide or minimize the latencies that result from slow memory access, such as data prefetching, memory partition, and multi-module architecture. But these approaches usually don't consider the optimization of energy consumption, which is critical to modern embedded computing systems. This chapter presents a novel loop scheduling and assignment algorithm with the combination of data prefetching, memory partition, and heterogenous assignment to obtain minimum energy consumption while hiding memory access latency.

Data prefetching, i.e., retrieving data from memory and storing it into the cache nearest to the CPU before using it, is an attractive approach to hide memory access latency [48, 49]. A prefetching scheme reduces the memory miss penalty by overlapping the processor computations with the memory access operations to achieve high throughput computation. Prefetching has been studied extensively. Previous research on prefetching can be classified into three categories: 1) prefetching based on hardware [58, 75, 202]; 2) prefetching based on software [131, 144]; 3) or both [51, 235].

Hardware-based prefetching techniques, requiring some support units connected to the cache, rely on the dynamic information available during program execution. In contrast, software prefetching schemes depend on compiler technology to analyze a program statically and insert explicit prefetching instructions into the program code. One advantage of software prefetching is that much compile-time information can be explored in order to effectively issue the prefetching instruction. Furthermore, many existing loop transformation approaches can be used to improve the performance of prefetching. Bianchini et al. [28] developed a runtime data prefetching strategy for software-based distributed shared-memory systems. Wallace and Bagherzadeh [217] proposed a mathematical model and a new prefetching mechanism. Mowry [138] proposed a new compiler algorithm for inserting prefetches into multiprocessor code. This algorithm attempts to minimize overheads by only issuing prefetches for references that are predicted to suffer cache misses.

To increase the data locality, one good method is to group the elemental computation points [147, 15, 102, 234]. Related work can be found in the loop tiling and the one-level partition techniques. Loop tiling is mainly applied to the distributed system to reduce communication time. Wolfe and Lam [230] proposed a loop transformation technique for maximizing parallelism or data locality. Bouilet et al. [31] introduced a criterion for defining optimal tiling in a scalable environment. In this method, an optimal tile shape can be determined and the tile size obtained from the resources constraints. Another interesting result was produced by Chame and Moon [39]. They proposed a new tile selection algorithm for eliminating self interference and simultaneously minimizing capacity and cross-interference misses. Their experimental results show that the algorithm consistently finds tiles that yield lower miss rates. Nevertheless, the traditional tiling techniques only concentrate on reducing communication cost. They do not consider how to best balance the computation and communication schedules. There is no energy optimization consideration in their algorithms also. Special cases occur in DSP and image processing applications, where there exists a large number of uniform nested loops. Thus, detailed consideration of how to schedule the loop body efficiently is very important to these applications.

Many scientific applications, such as multimedia, DSP and image processing, usually contain repetitive groups of operations represented by nested loops with uniform data dependencies. These applications can be modeled by *multi-dimensional data flow graphs* (MDFGs) [151]. Chen et al. [49, 48] develop a methodology which combines loop pipelining and data prefetching. The algorithm investigates the inter-iteration dependencies when considering data prefetching. It produces two schedules: ALU computation part and memory access (prefetching) part, and also reduces the overall schedule length by balancing the lengths of these two schedules.

Much research has been conducted in the area of using multi-module memory to achieve maximum instruction level parallelism, i.e., optimize performance [117, 198, 54, 123, 179, 232]. These approaches differ in either the models or the heuristics. However, none of these works consider the combined effect of performance and energy requirements. Actually, performance requirement often conflicts with energy saving [60, 112, 213, 61, 150, 238]. Hence, significant energy saving and performance improvements can be obtained by exploiting heterogeneous multi-module memory at the instruction level. Wang et al. [222] have considered the combined effect and proposed the VPIS algorithm to overcome it, but their algorithm only considers part of the scenario. They haven't considered heterogeneous memory module type assignment, data prefetching, and memory partition.

Combining the consideration of energy and performance, in this chapter, we propose a novel architecture model to overcome the weakness of previous techniques. We designed an algorithm, LSM (*Loop Scheduling for Multicore*), to minimize total energy while hiding memory latency for loop applications. In the LSM algorithm, the schedules are generated by repeatedly rotating down and reallocating nodes with shorter schedule length based on rotation scheduling [41]. After the best schedule that has the balanced schedule length for processor core and memory is selected, we use memory partition with data prefetching, variable partition, and heterogeneous module type assignment to obtain the minimum total energy while hiding memory latency.

The experimental results show that LSM achieves a significant reduction on average in total energy consumption. For example, with three module types, compared with the VPIS in [222], LSM shows an average 21.0% reduction in total energy consumption while satisfying timing constraint 300. Compared with algorithms in paper [222], our algorithm has two advantages: First, our algorithm can find better solutions that result in smaller total energy consumption while satisfying timing constraints. Second, it provides a promising approach to hide memory latency for applications with nested loops.

In summary, we have three major contributions in this chapter: First, we study the combined effects of energy and performance of memory and processing core in a systematic approach. Second, we exploit memory energy saving with loop scheduling and heterogeneous memory module type assignment. Third, performance has been increased compared with the scenario without using our techniques. We use prefetching and keep operations to hide the memory latencies. Memory size usage has been optimized by using memory partition, variable partition and loop scheduling.

In the next section, we introduce the necessary background, including basic definition and models. The algorithm is discussed in Section 3.3. Experimental results are provided in Section 3.4.

3.2 Basic Concepts and Models

In this section, we introduce some basic concepts which will be used in the later sections. First the architecture model is given. Next, we introduce the multi-module, two-mode memory model. Then we introduce the MDFG (multi-dimensional data-flow graph) model and rotation scheduling. Partitioning of iteration space is presented next. Finally, we introduce type assignment with a detailed example.

3.2.1 The Architecture Model

Our technique is designed for use in a system containing processors with multiple processing cores and memory modules (banks), as shown in Figure 3.1. The on-chip memory has a tight memory size constraint and fast access velocity while the remote memory (with multi-port) is larger and slower. This architecture is similar to the real system with L1 cache, L2 cache, and main memory. Our approach is to load data into the on-chip memory first and wait for the processing core. In this way, the overall cost of accessing data can be minimized. By overlapping the core computations and the memory accesses, we can achieve a shorter overall execution time. The key idea of our algorithm is to tolerate the memory access time by overlapping the core computations as much as possible.

Data Flow Graph (DFG) is used to model many multimedia and DSP applications. The definition is as follows:

Definition 3.1 *A DFG $G = \langle U, ED, T, E\rangle$ is a node-weighted directed acyclic graph (DAG), where $U = \langle u_1, \cdots, u_i, \cdots, u_N\rangle$ is a set of operation nodes; $ED \subseteq U \times U$ is an edge set that defines the precedence relations among nodes in U; T is a set of operation time for all nodes in U; E is a set of energy consumption for all nodes in U.*

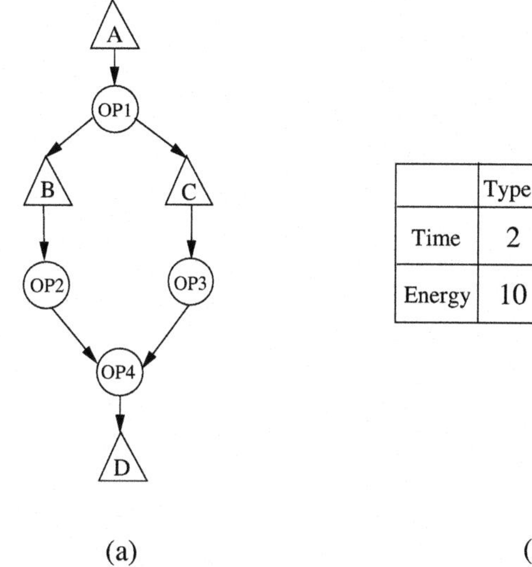

(a)

	Type1	Type2	Type3
Time	2	3	5
Energy	10	3	1

(b)

FIGURE 3.2: (a) A DFG with both core and memory operations. (b) The types of memory.

The nodes in the data flow graph can be either memory operations or core operations. As shown in Figure 3.2(a), the core operations are represented by circles, and the memory operations are represented by triangles. Particularly, an edge from a memory operation node to a core operation node represents a *Load* operation, whereas the edge from a core operation node to a memory operation node represents a *Store* operation. The memory node with same alphabet prefix but different number postfix stands for the same variable but different memory access operations. For instance, node $A1$ and $A2$ are memory operations accessed by the same variable A, but they are different memory operations.

Figure 3.2(a) shows a DFG. We start from loading value of variable A into core to do operation 1, i.e., $OP1$. Then load variable B to do $OP2$. The same is for $OP3$. Next, do $OP4$ with the inputs of the results of $OP2$ and $OP3$. Finally, store the result of $OP4$ into variable A in memory. In this example, we have two same type cores; they can finish each operation in 1 clock cycle with energy consumption 1 μJ. Assume there are three types of banks to be chosen from and we can only use maximum two types of banks. Only two banks can be accessed at the same time. The memory types are shown in Figure 3.2(b). Type 1 has memory access time latency 2 clock cycles with energy consumption 4 μJ; Type 2 has memory access time latency 3 with energy consumption 3; and the time is 5 and energy is 1 for type 3. We can represent them in Type(Time, Energy) format, such as 1(2, 4). There is a timing constraint L and it must be satisfied for executing the whole DFG, including both memory access part and core part.

In nested loop applications, an *iteration* is the execution of all the operations of a loop one time. We divided each iteration space into partitions, which will be discussed in detail in later subsections. Our prefetching scheme is a software-based method in which some special prefetching instructions are added to the code when it is compiled [224, 48, 49, 223]. When the processor encounters these instructions during program execution, it will pass them to the SHMU (special hardware memory units) for handling. The function of the memory unit is: get the data ready before the on-chip memory needs to reference them. Two types of instructions, *prefetch* and *keep*, are supported by memory units. The keep instruction keeps the data in the memory for use during a later partition's execution, which will be described in later subsections. In the on-chip memory, depending on the partition size and different delay dependencies, the data will need to be kept for different amounts of time. The advantage of using keep is that it can eliminate the time wasted for unnecessary data swapping. If a delay dependence starts from an iteration in the current partition and terminates in the next partition, a memory operation is used to keep this data in the on-chip memory for the time of partition executions. Delay dependencies that go into other partitions result in the use of prefetch memory operations to fetch data in advance. The relation of processing core and memory in a partition is shown in Figure 3.3.

To form the processing core schedule, a multidimensional rotation scheduling algorithm [151] is used to obtain the schedule of a single iteration. Then, this schedule is simply duplicated for each iteration in the partition. The memory schedule is formed by considering all the memory operations in the entire partition. The partition size is selected to balance the core and memory schedules such that the long memory latency is effectively tolerated.

3.2.2 The Memory Model

To improve the overall performance, many DSPs employ a Harvard architecture, which provides simultaneous accesses to separate on-chip memory modules for instructions and data [137, 13, 204, 89]. Some DSP processors are further equipped with multi-module memory that is accessible in parallel, such as Analog Device ADSP2100, Motorola DSP56000, NEC uPd77016, and Gepard Core DSPs [222, 137, 89, 117]. Harvesting the benefits provided by the multi-module memory architecture hinges on sufficient compiler support. Parallel operations afforded by multi-module memory give rise to the problem of how to maximally utilize the instruction level parallelism.

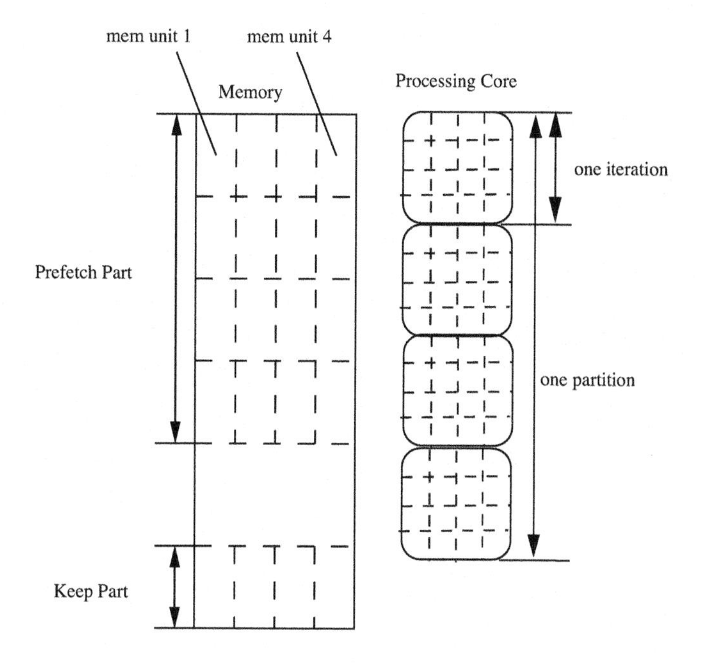

FIGURE 3.3: Processing core and memory in a partition.

In multi-module memory, each module can adjust its operating mode independently. One example of this type of memory architecture is RDRAM [175], in which each module can work with four operating modes: active, standby, nap, and power-down. Our model has two modes: active mode and inactive mode. A module can service a read/write request in the active mode [60]. If it is not servicing a memory request, it can be placed into the inactive operating mode to save power. A major problem with this arrangement is that it incurs an additional resynchronization penalty when a module in this mode needs to service a memory request. Consequently, it is advantageous to keep memory modules in idle conditions for long durations of time so that the resynchronization penalties can be compensated. We achieve this by clustering data accesses in a small number of memory modules and placing the remaining idle modules into the inactive operating mode. Variable partition is an important method to improve the data locality. Different variable partitions will significantly affect the schedule length and the energy consumption of an application. The details of variable partition and scheduling algorithm are described in [222]. The operating mode transition is controlled by the memory controller, whose states can be modified through a set of configuration registers [61].

3.2.3 MDFG and Rotation Scheduling

Multi-dimensional Data-Flow Graph (MDFG) is used to model the nested loops of an application. A *MDFG* $G = \langle U, ED, d \rangle$ is a node-weighted and edge-weighted directed graph, where $U = \langle u_1, \cdots, u_i, \cdots, u_N \rangle$ is the set of operation nodes, $ED \subseteq U \times U$ is the edge set that defines the precedence relations among nodes in U, and $d(ed)$ is a function from ED to Z^n, representing the number of delays for an edge *ed*. Figure 3.4(a) shows the original MDFG of Wave Digital Filter (WDF).

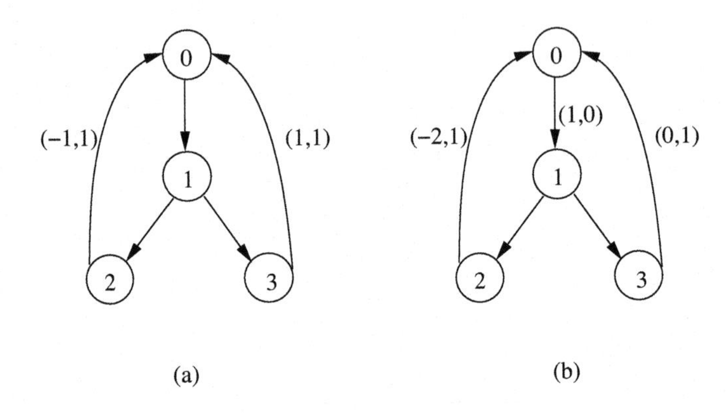

FIGURE 3.4: (a) The original MDFG of Wave Digital Filter. (b) The retimed MDFG after rotating node 0.

An *iteration* is the execution of each node in U once. Iterations are identified by a vector $\bar{\beta}$. Inter-iteration dependencies are represented by vector-weighted edges. For any iteration $\bar{\beta}$, an edge *ed* from u to v ($u \rightarrow v$) with delay vector $d(ed)$ means that the computation of node v at iteration $\bar{\beta}$ depends on the execution of node u at iteration $\bar{\beta} - d(ed)$. An edge with zero-delay (0, $\cdots$, 0) represents a data dependence within the same iteration. A legal MDFG must not have zero-delay cycles. A *legal* MDFG G $= \langle U, ED, d \rangle$ is realizable if there exists a *schedule vector* s for the MDFG, such that $s \cdot d(ed) \geq 0$ for any $ed \in ED$ [152]. A schedule vector s is the normal vector for a set of parallel equitemporal hyperplanes. Iterations in the same hyperplane will be executed in sequence. Iterations are represented as integral points in a Cartesian space, called *iteration space*, where the coordinates are defined by the loop control indexes.

A *multi-dimensional (MD) retiming* r is a function from U to Z^n that redistributes the nodes while producing a new MDFG G_r. $r(u)$ represents delay components pushed into all outgoing edges $u \rightarrow v_i$; and subtracted from all incoming edges $w_j \rightarrow u$, where $u, v_i, w_j \in G$. Therefore, we have $d_r(ed) = d(de) + r(u) - r(v)$ for every edge $u \rightarrow v$. The legal retiming vector r can be any vector which is orthogonal to the scheduling vector s. In multi-dimensional retiming algorithms, r is chosen as the base vector which is orthogonal to s.

Multi-dimensional rotational scheduling is a loop pipelining technique which implicitly uses MD retiming heuristic for scheduling cyclic graphs. In this algorithm, those nodes with all the incoming delay edges different from

$(0, \cdots, 0)$ are called *rotatable nodes.* At each step, the rotatable nodes from the first schedule row are rotated, combined with pushing a retiming vector $r = (r_x, r_y)$ through them. By doing this, the data dependencies between these rotatable nodes and other nodes are transformed. Then those rotated nodes will be partially rescheduled to earlier control steps. If all the rotated nodes have been moved up, the schedule length can be reduced. Given an initial schedule, the rotation technique repeatedly transforms the schedule to a more compact one under the resource constraint. Figure 3.4(b) shows the retimed MDFG after rotating node 0.

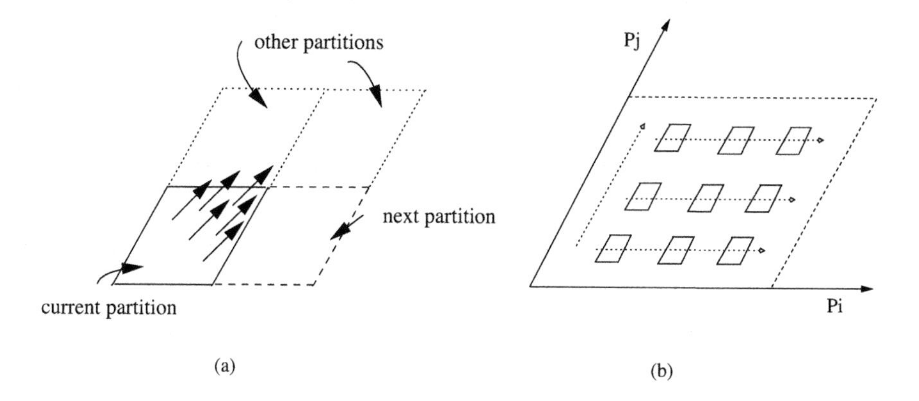

FIGURE 3.5: (a) Partitions: solid edges represent *prefetch* operations; dashed edges represent *keep* operations; dotted edges are the data dependencies inside the partition, hence no memory operation is needed. (b) Iterations will be executed from left to right in the P_i direction and then proceed to the next hyperplane along the direction perpendicular to P_i.

3.2.4 Partitioning the Iteration Space

Our scheduling consists of two parts, the processing core part and the memory part. The core part of the schedule for one iteration is generated by applying the rotation scheduling algorithm. Rotation scheduling is a loop pipelining technique which implicitly uses retiming heuristic for scheduling cyclic graphs.

Scheduling of the memory part consists of several steps. First, we need to decide a legal partition. Second, partition size is calculated to ensure an optimal schedule under no memory constraint. Third, memory size of the partition is calculated to see if it fits the memory constraint, and the partition size is updated accordingly to make sure the memory constraint is met. Fourth, iterations are numbered and both the processing core part and memory part of the schedule are generated.

Regular execution of nested loops proceeds in either a row-wise or column-wise manner until the end of the row or column is reached. However, this mode of execution does not take full advantage of either the reference locality or the available parallelism since dependencies have both horizontal and vertical components. The execution of such structures can be made to be more efficient by dividing the iteration space into regions called *partitions* that better exploit spatial locality. Provided that the total iteration space is divided into partitions, the execution sequence will be determined by each partition. Partitions are executed from left to right. Within each partition, iterations are executed in row-wise order. At the end of a row of partitions, we move up to the next row and continue from the far left in the same manner. In the two-level partition approach, the iteration space will be partitioned on two levels. The first-level partition consists of a number of iterations, and the second-level partition is made up of a number of first-level partitions.

Assume that the partition in which the loop is executing is the *current partition*. Then, the *next partition* is the partition adjacent on the right side of the current partition along the horizontal axis. The second next partition is adjacent to and lies on the right side of the next partition, with the definitions of third next partition, fourth next partition, etc., similar. The *other partitions* are all partitions except for those on the current row. The different partitions are shown in Figure 3.5(a).

3.2.4.1 Partition Size

A partition is identified by two partition vectors, P_i and P_j, where $P_i = P_{i0} \times f_i$ and $P_j = P_{j0} \times f_j$. While P_{i0} and P_{j0} determine the direction and shape of a partition, f_i and f_j determine the size of a partition. In Figure 3.5(b), the outside parallelogram represents a partition. The iterations are numbered from left to right in the P_i direction, then to the next hyperplane along with the direction of the vector perpendicular to P_i. The small parallelograms inside represent the iterations in the partition.

3.2.4.2 The Memory Size

The on-chip memory should be large enough to hold all the data which are needed during the execution of the partitions. We classify the on-chip memory requirements into three categories: basic memory for the working set, reserved memory for prefetch operations and reserved memory for keep operations. The first corresponds to all the internal delay edges inside the current partition. For the second and third categories, reserved memory for prefetch and keep operations, these memory operations represent the data instances pre-loaded or pre-stored in the on-chip memory before we execute each partition. One memory location should be reserved for each one of these memory operations. The total number of this pre-stored data is two times the total number of memory operations (one for the pre-loaded data of the current partition; the other for the newly generated data for the next partition). Therefore, the size of this part of memory is: $Size_{reserved} = 2(NUM_{prefetch} + NUM_{keep})$. Finally, the on-chip memory needed to execute this partition is: $On\_Chip\_Size = Size_{basic} + Size_{reserved}$. This estimation gives the designer a good indication of how much on-chip memory is required [49, 48].

3.2.5 Type Assignment

3.2.5.1 Type Assignment for Memory Only

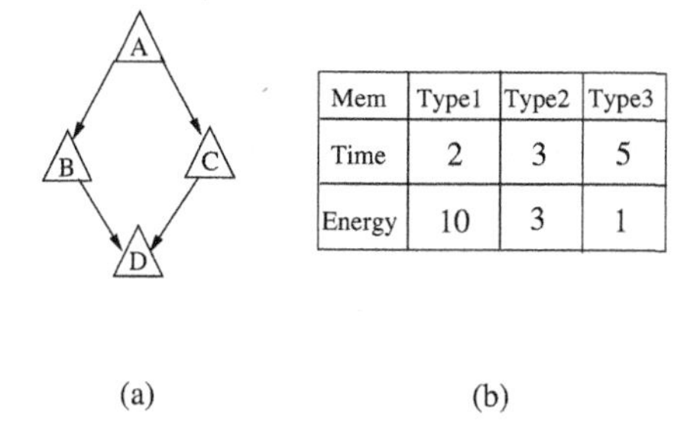

Mem	Type1	Type2	Type3
Time	2	3	5
Energy	10	3	1

FIGURE 3.6: (a) DFG with four memory operations. (b) The types of memory modules.

Assume we have a task graph for memory, which is represented by a *data flow graph* (DFG) and is shown in Figure 3.6. Assume only two modules can be accessed at the same time and three types of modules are available. Therefore, we can use only maximum two types of modules. The memory types are shown in Figure 3.6(b). Type 1 has memory access time latency 2 clock cycles with energy consumption 10 μJ, type 2 has memory access time latency 3 clock cycles with energy consumption 3 μJ, and the time is 5 clock cycles and energy consumption is 1 μJ for type 3. We can represent them in Type (Time, Energy) format, such as type 1 (2, 10).

Assume after implementing variable partition [222] on this DFG, we know that B and C need to be assigned into different modules. We can implement type assignment to minimize total energy while satisfying different timing constraints L to hide memory latency. For instance, under timing constraint 8, the type assignment for memory operations is: A, B in one module with type 1 (2, 10) and C, D in another module with type 2 (3, 3). The total time for memory part is 8, and the total energy consumption is 26. The detailed schedule is shown in Figure 3.7(a). If we don't have multiple modules to select (only one memory type is allowed), i.e., for the homogeneous situation, we can choose type 1 (2, 10) since the total time cannot satisfy timing constraint 8 with type 2 or 3. The detailed schedule with homogeneous memory (type 1 (2, 10)) is shown in Figure 3.7(b). The total energy consumption is 40. The heterogeneous memory type solution consumes just 65% energy of homogeneous memory type solution.

By using our *Type_Assign* algorithm, we obtained the minimum energy

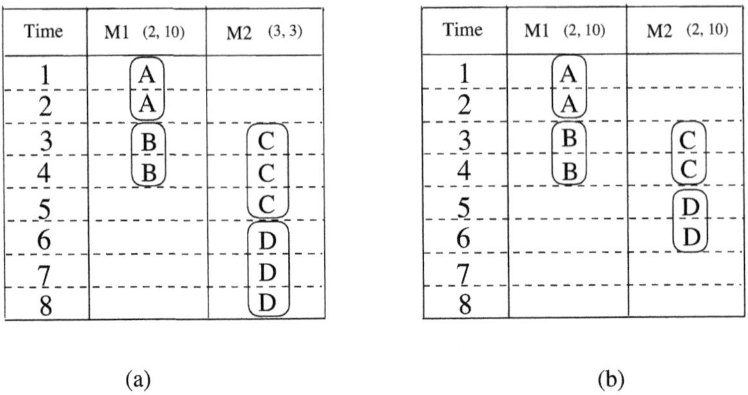

Time	M1 (2, 10)	M2 (3, 3)
1	A	
2	A	
3	B	C
4	B	C
5		C
6		D
7		D
8		D

(a)

Time	M1 (2, 10)	M2 (2, 10)
1	A	
2	A	
3	B	C
4	B	C
5		D
6		D
7		
8		

(b)

FIGURE 3.7: (a) The best schedule of *Type_Assign* using two memory types with timing constraint 8. (b) The best schedule using only one memory type with timing constraint 8.

consumptions satisfying different timing constraints for Figure 3.6. The results are shown in Table 3.1. "Time" represents total time spent and "Energy" represents total energy consumption of the DFG. The whole solution space for this DFG only has seven non-repetitive solutions with different timing constraints.

Time	**6**	**7**	**8**	**9**	**11**	**13**	**15**
Energy	40	33	26	12	10	8	4

TABLE 3.1: The solution table of the DFG in Figure 3.6(a) by using *Type_Assign*.

3.2.5.2 Type Assignment for Core and Memory Together

After the above simple example for memory part only, now we continue to solve the example in Figure 3.2(a) and give the final solution table by using our type assignment method.

From variable partition, we know B and C should be in different banks. Also A and D should be in same bank, i.e., different operations corresponding to same variable. Based on precedence relations in Figure 3.2(a) and variable partition information, we do type assignment to minimize total energy under different timing constraints L. For instance, under timing constraint 10, the type assignment for memory operations is: A, B, D in one bank with type 1(2, 10) and C in another bank with type 2(3, 3). The total time for memory part is 7, and the total energy consumption for memory part is 33. In this example, the DFG of memory is shown in Figure 3.8(a). Adding up the ALU part, which have total time 3 and total energy consumption 4, we get the final result: the total time is 10 and the total energy consumption is 37. The result is shown in Figure 3.8(b).

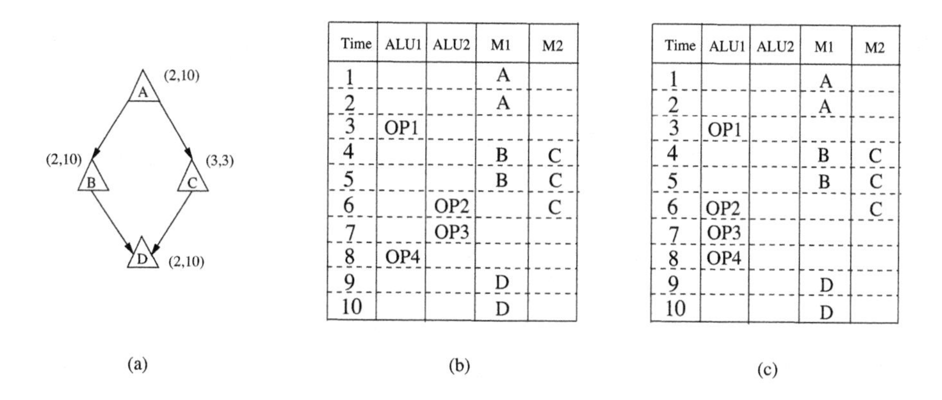

FIGURE 3.8: (a) The memory DFG. (b) The result of *Type_Assign* for the DFG in Figure 3.2(a) with timing constraint 10. (c) The minumum resource scheduling for the DFG in Figure 3.2(a) with timing constraint 10.

After the type assignment, we implement minimum resource scheduling. For example, at timing constraint 10, we only need one ALU and two memories. One memory is type 1, and the other is type 2. In Figure 3.8(b), we can move operations $OP2$ and $OP3$ to column ALU1 and let ALU1 execute them. The result is shown in Figure 3.8(c)

Based on Figure 3.2(a), we obtained the minimum energy consumptions satisfying different timing constraints by using our algorithm. The results are shown in Table 3.2. "T" represents total time spent and "E" represents total energy consumption of the DFG. There are only total five solutions with different timing constraints based on the restrictions of different inputs.

Time	**9**	**10**	**11**	**12**	**14**	**16**	**18**
Energy	44	37	30	16	14	12	8

TABLE 3.2: The results of *Type_Assign* for the DFG in Figure 3.2(a).

3.3 The Algorithms

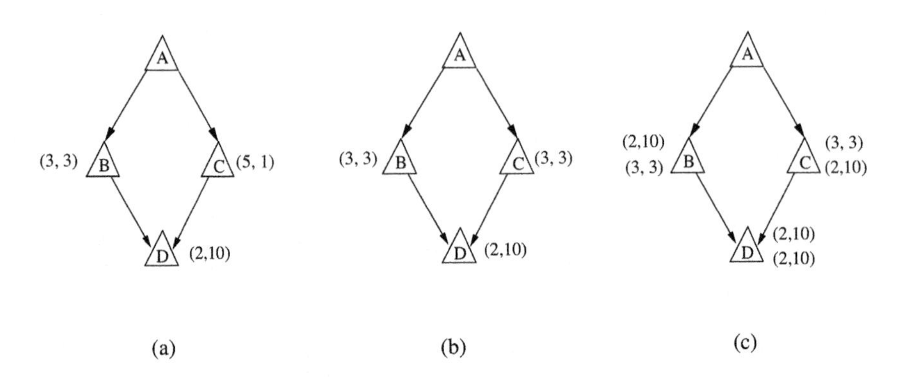

FIGURE 3.9: (a) Example of *Rule 1.* (b) Example of *Rule 2.* (c) Example of *Rule 3.*

In this section, an algorithm, LSM (*Loop Scheduling for Multicore*), is designed to solve the problem of minimizing total energy without sacrificing performance based on loop scheduling. The basic idea is to generate the schedules by repeatedly rotating down and reallocating nodes with minimum total energy consumption based on Rotation Scheduling, and then select a best schedule and assignment that has the minimum total energy. During each iteration, we implement processing core scheduling and memory scheduling.

3.3.1 The LSM Algorithm

The LSM algorithm is shown in Algorithm 3.3.1. In this algorithm, we first put all nodes in the first row of S into set U_r. Then we delete the first row of S and shift S up by one control step. After that, we retime each node $u \in U_r$ such that $r(u) \leftarrow r(u) + 1$. Then based on the precedence relation in the retimed graph G_r, we rotate each node $u \in U_r$ by putting u into the earliest location. Next, for each new schedule, we do processing core scheduling and memory scheduling simultaneously using the algorithm in [48, 49]. LSM will repeat the above procedure R times, where R is a user-specified amount. The best (balanced) schedule S_j is selected from the generated R schedules. Then we implement memory partition with data prefetching and variable partition. Finally, use algorithm *Type_Assign* on schedule S_j to obtain the minimum total energy E. The key idea of algorithm *Type_Assign* is using dynamic programming to assign memory type for each memory operation.

LSM algorithm has several advantages: 1) Using loop scheduling to exploit the pipelining and regular pattern of an application. 2) Hiding the memory latencies with prefetching and keep operations. 3) Balancing the core scheduling

---

Algorithm 3.3.1 LSM

Require: MDFG $G = \langle U, ED, d \rangle$,
the retiming r of G,
an initial schedule S of G,
the rotation times R,
M different memory module,
P different processing cores,
the timing constraint L.
Ensure: A schedule S, an assignment A, and the retiming r, to MIN(E)

1: **for all** $i = 1$ to R **do**
2: $U_r \leftarrow$ All nodes in the first row in S;
3: Delete the first row from S;
4: Shift S up by 1 control step;
5: **for all** $u \in U_r$ **do**
6: $r(u) \leftarrow r(u) + 1$;
7: **end for**
8: Using revised algorithm in [48] for core scheduling ;
9: Using revised algorithm in [49] for memory scheduling ;
10: Using overlapping to balance the scheduling length of core and memory schedules;
11: **end for**
12: $S_j \leftarrow$ the best schedule S; $r_j \leftarrow$ corresponding r;
13: Using revised algorithm in [49] for memory partition and data prefetching;
14: Using VPIS in [222] for variable partition and mode scheduling;
15: Using algorithm *Type_Assign* to obtain MIN(E) module type assignment A for the schedule S_j;
16: Using *Minimum Resource Scheduling and Configuration* algorithm to implement processing core and memory scheduling;
17: $E_{min} \leftarrow$ MIN(E);
18: Store the S_j, r_j, and A corresponding to E_{min};
19: Output S_j, r_j, and A

Algorithm 3.3.2 Algorithm to Compute $D_{i,j}$ for a Simple Path

Require: A simple path DFG
Ensure: $D_{i,j}$
1: Build a local table $B_{i,j}$ for each node;
2: Start from u_1, $D_{1,j} \leftarrow B_{i,j}$;
3: **for all** $u_i, i > 1$, **do**
4:    **for all** timing constraint j, **do**
5:       Compute the entry $D_{i,j}$ as follows:
6:       **for all** k in $B_{i,j}$, **do**
7:          $D_{i,j} = D_{i-1,j-k} + B_{i,k}$;
8:          Cancel redundant energy and linked list with the three redundant linked list cancellation rules;
9:          Insert $D_{i,j-1}$ to $D_{i,j}$ and remove redundant energy value according to Lemma 3.3.1;
10:       **end for**
11:    **end for**
12: **end for**

and memory scheduling to compact the total schedule length. 4) Optimizing the memory size requirement and exploiting spatial locality by partition. 5) Using variable partition to exploit the spatial locality and save power. 6) Using mode scheduling, i.e., exploit the inactive mode to save power by clustering data accesses in a small number of memory modules and placing the remaining idle modules into the inactive operating mode. 7) Exploiting the heterogeneous multi-module memory architecture by using novel type assignment to save energy.

3.3.2 The Type_Assign Algorithm

3.3.2.1 Definitions and Lemma of Type_Assign

To solve the type assignment problem for a certain schedule S, we use a dynamic programming method traveling the graph in a bottom up or top down fashion. For the ease of explanation, we will index the nodes based on bottom up sequence.

Given the timing constraint L, a DFG G, and an assignment A, we first give several definitions as follows:

1. The function from domain of variable to range of module-type is defined as Module(). For example, "Module(A)= type 1" means the module-type of variable A is type 1.

2. G^i: The sub-graph rooted at node u_i, containing all the nodes reached by node u_i. In our algorithm, each step will add one node which becomes the root of its sub-graph.

3. $E_A(G^i)$ and $T_A(G^i)$: The total energy consumption and total execution time of G^i under the assignment A. In our algorithm, each step will achieve the minimum total energy consumption of G^i under various timing constraints.

4. In our algorithm, table $D_{i,j}$ will be built. Here, i represents a node number, and j represents a timing constraint. Each entry of table $D_{i,j}$ will store energy consumption $E_{i,j}$ and its corresponding linked list. Here we define $E_{i,j}$ as follows: $E_{i,j}$ is the minimum energy consumption of $E_A(G^i)$ computed by all assignments A satisfying $T_A(G^i) \leq j$. The linked list records the type selection of all previous nodes passed, from which we can trace back how $E_{i,j}$ is obtained.

We have the following lemma about energy consumption cancellation with the same timing constraint.

Lemma 3.3.1 *Given* $E^1_{i,j}$ *and* $E^2_{i,j}$ *with the same timing constraint, if* $E^1_{i,j} \geq E^2_{i,j}$, *then* $E^2_{i,j}$ *will be kept.*

In each step of dynamic programming, we have several rules about cancellation of redundant energy consumption and its corresponding linked list.

1. **Rule 1:** If the number of memory types is greater than K, which is the maximum number of modules that can be accessed simultaneously, discard the corresponding energy consumption and corresponding linked list.

2. **Rule 2:** If two siblings a and b, i.e., children of same node, are not allowed to be same type, i.e., Module(a) $\neq$ Module(b), in variable partition, and Module(a) = Module(b) in assignment, then discard the corresponding E and corresponding linked list, except the scenario that all the nodes till now are in the same type, i.e., $\forall u, v \in G$, Module(u) = Module(v).

3. **Rule 3:** If two siblings a and b are just exchanging their types and other nodes are the same in type assignment for the two corresponding linked lists, i.e., Module(a) = Module(b), then only keep one E and its corresponding linked list.

For example, in Figure 3.9(a), using bottom up approach, in the fourth step, i.e., for node A, if we get a sequence (2,10)(3,3)(5,1) for D, B, C nodes, then discard this sequence by *Rule 1*. In this case, three memory types are required, but there are only two memory types available. In Figure 3.9(b), B and C are not allowed to be the same type except all A, B, and C are the same type. If we get a sequence (2,10)(3,3)(3,3), then delete it by *Rule 2*. But we will keep a sequence if it is (3,3)(3,3)(3,3). In Figure 3.9(c), at node A, if we get two sequences (2,10)(3,3)(2,10) and (2,10)(2,10)(3,3), then based on *Rule 3* (symmetric rule), we only keep one sequence.

In every step of our algorithm, one more node will be included for consideration. The data of this node is stored in local table $B_{i,j}$, which is similar to table $D_{i,j}$, but with energy consumption only on node u_i. A local table stores only data of energy consumption of a node itself. Table $B_{i,j}$ is the local table only storing the energy consumption of node u_i. $E_{i,j}$ is the energy consumption only for node u_i with timing constraint j. The algorithm to compute $D_{i,j}$ is shown in Algorithm 3.3.2.

3.3.2.2 The Type_Assign Algorithm

In algorithm *Type_Assign*, which is shown in Algorithm 3.3.3, without loss of generality, assume using bottom up approach. Algorithm *Type_Assign* gives the near-optimal solution when the given DFG is a DAG. In step 6, $D_{i_1,j} + D_{i_2,j}$ is computed as follows. Let G' be the union of all nodes in the graphs rooted at nodes u_{i_1} and u_{i_2}. Travel all the graphs rooted at nodes u_{i_1} and u_{i_2}. If a node q in G' appears for the first time, we add the energy consumption of q to $D'_{i,j}$. If q appears more than once, that is, q is a common node, we only count it once. That is, the energy consumption is just added once. Then use a selection function to choose the type of q. For instance, the selection function can be defined as selecting the type that has a smaller time. Since each node can only have one assignment, there is no assignment conflict. From step 7, the minimum total energy consumption is selected from all possible energy consumption caused by adding u_{i+1} into the sub-graph rooted on u_{i+1}. So for each $E_{i+1,j}$ in $D_{i+1,j}$, $E_{i+1,j}$ is the total energy consumption of the graph G^{i+1} under timing constraint j.

3.3.2.3 The Minimum Resource Scheduling and Configuration

We have obtained type assignment with at most K types of memory banks and P types of core. Then we will compute the configuration of the types and give the corresponding schedule which consumes minimum energy while satisfying timing constraint. We propose minimum resource scheduling algorithms to generate a schedule and a configuration which satisfies our requirements. We first propose *Algorithm Lower_Bound_RC* that produces an initial configuration with low bound resource. Then we propose *Algorithm Min_RC_Scheduling* that refines the initial configuration and generates a schedule to satisfy the timing constraint.

Algorithm Lower_Bound_RC is shown in Figure 3.10. In the algorithm, it counts the total number of every type in every time unit in the ASAP and ALAP schedule. Then the lower bound for each bank type is obtained by the maximum value that is selected from the average resource needed in each time period. For example, for the DFG in Figure 3.2(a), after using *Min_RC_Scheduling* algorithm, we find that the lower bound of both core and memory are 1.

Using the lower bound of each type as an initial configuration, we propose an algorithm, *Algorithm Min_RC_Scheduling*, which is shown in Figure 3.11,

---

Algorithm 3.3.3 *Type_Assign* Algorithm

Require: DFG $G = \langle U, ED, T, E \rangle$ with N nodes, M types of memory modules with (T, E) pairs, K number of memory modules that can be accessed simultaneously, P numbers of the processing cores, the timing constraint L

Ensure: An efficient type assignment to MIN(E) while satisfying L

1. $SEQ \leftarrow$ Sequence obtained by topological sorting all the nodes.

2. $t_{mp} \leftarrow$ the number of multi-parent nodes; $t_{mc} \leftarrow$ the number of multi-child nodes;
   If $t_{mp} < t_{mc}$, **Then** use bottom up approach;
   Else, use top down approach.

3. **If** bottom up approach, **Then** use the following algorithm;

   If top down approach, **Then** just reverse the sequence.

4. $SEQ \leftarrow \{u_1 \rightarrow u_2 \rightarrow \cdots \rightarrow u_N\}$, in bottom up fashion;
   $D_{1,j} \leftarrow B_{1,j}$;
   $D'_{i,j} \leftarrow$ the table that stored MIN(E) for the sub-graph rooted on u_i except u_i;
   $u_{i_1}, u_{i_2}, \cdots, u_{iW} \leftarrow$ all child nodes of node u_i; $w \leftarrow$ the number of child nodes of node u_i.

5. **If** w = 0, **Then** $D'_{i,j} = (0, 0)$;
   If w = 1, **Then** $D'_{i,j} = D_{i_1,j}$;
   If w > 1, **Then** $D'_{i,j} = D_{i_1,j} + \cdots + D_{i_w,j}$;

6. Computing $D_{i_1,j} + D_{i_2,j}$:
   $G' \leftarrow$ the union of all nodes in the graphs rooted at nodes u_{i_1} and u_{i_2};
   Travel all the graphs rooted at nodes u_{i_1} and u_{i_2};
   If a node is a common node, **Then** use a selection function to choose the type of a node.
   Cancel redundant energy and linked list with the three redundant linked list cancellation rules.

7. For each k in $B_{i,k}$,
   $D_{i,j} = D'_{i,j-k} + B_{i,k}$

8. $D_{N,j} \leftarrow$ a table of MIN(E);
   Output $D_{N,L}$

Input: A DFG with type assignments and timing constraint L
Output: Lower bound for each type
Algorithm:

1. Schedule the DFG by ASAP and ALAP scheduling, respectively.
2. $N_{ASAP}[i][j] \leftarrow$ the total number of nodes with type j and scheduled in step i in the ASAP schedule.
3. $N_{ALAP}[i][j] \leftarrow$ the total number of nodes with type j and scheduled in step i in the ALAP schedule.
4. For each type j,
$$LB_{ASAP}[j] \leftarrow \max\{N_{ASAP}[1][j]/1,$$
$$(N_{ASAP}[1][j] + N_{ASAP}[2][j])/2,$$
$$\cdots, \sum_{1 \le k \le L} N_{ASAP}[k][j]/L\}$$
5. For each type j,
$$LB_{ALAP}[j] \leftarrow \max\{N_{ALAP}[L][j]/1,$$
$$(N_{ALAP}[L][j] + N_{ALAP}[L-1][j])/2,$$
$$\cdots, \sum_{1 \le k \le L} N_{ALAP}[L-k+1][j]/L\}$$
6. For each type j, its lower bound:
$$LB[j] \leftarrow \max\{LB_{ASAP}[j], LB_{ALAP}[j]\}$$

FIGURE 3.10: Algorithm Lower_Bound_RC.

Input: A DFG with type assignments, timing constraint L, and an initial configuration
Output: A schedule and a configuration
Algorithm:

1. For each node v in DFG, compute ALAP(v) that is the schedule step of v in the ALAP schedule.
2. $S \leftarrow 1$.
3. Do {
    - Ready_List $\leftarrow$ all ready nodes;
    - For each node $v \in$ Ready_List, if ALAP(v)==S, schedule v in step S with additional resource if necessary;
    - For each node $v \in$ Ready_list, schedule node v without increasing current resource or schedule length;
    - Update Ready_List and $S \leftarrow S + 1$;

   } **While** ($S \leq L$);

FIGURE 3.11: Algorithm Min_RC_Scheduling.

to generate a schedule that satisfies the timing constraint and gets the final configuration. In the algorithm, we first compute ALAP(v) for each node v, where ALAP(v) is the schedule step of v in the ALAP schedule. Then we use a revised list scheduling to perform scheduling. In each scheduling step, we first schedule all nodes that have reached to the deadline with additional resource if necessary and then schedule all other nodes as many as possible without increasing resource. For example, for the DFG in Figure 3.2(a), after using *Min_RC_Scheduling* algorithm, we find that only one core and two memory (one is type 1, the other is type 2) are needed. The scheduling is shown in Figure 3.2(b).

Algorithms *Lower_Bound_RC* and *Min_RC_Scheduling* both take $O(|U| + |ED|)$ to get results, where $|U|$ is the number of nodes and $|ED|$ is the number of edges for a given DFG.

3.4 Experiments

In this section, we implement our LSM algorithm in the SPAM compiler environment [198]. We conduct experiments with our algorithm on a set of benchmarks including Wave Digital filter (WDF), Infinite Impulse filter (IIR), Differential Pulse-Code Modulation device (DPCM), two dimensional filter

(2D), Floyd-Steinberg algorithm (Floyd), and All-pole filter. The proposed runtime system has been implemented and a simulation framework to evaluate its effectiveness has been built. The distribution of execution times of each node of DFG is Gaussian. Five memory modules are used in our experiments. w different types, $B_1, \cdots, B_w$, are used in the system, in which type B_1 is the fastest with the highest energy consumption and type B_w is the slowest with the lowest energy consumption.

We conducted experiments on five methods: Method 1: list scheduling, without memory partition, data prefetching, and type assignment; Method 2: the VPIS in [222], without memory partition, data prefetching, and type assignment; Method 3: list scheduling plus memory partition and data prefetching. Method 4: the VPIS in [222] plus memory partition and data prefetching; Method 5: our LSM algorithm. In the list scheduling, the priority of a node is set as the longest path from this node to a leaf node [134]. The experiments are performed on a PC with a P4 2.1 G processor and 512 MB memory running Red Hat Linux 9.0. In the experiments, the running time of LSM on each benchmark is less than one minute.

The experimental results for the five methods are shown in Table 3.3 to Table 3.5 when the number of module types is (3, 4, 5) and the corresponding module numbers that can be accessed simultaneously is (3, 4, 5). Column "Bench." represents the types of benchmarks and column "Num." represents the number of nodes of each filter benchmark. Column "E" represents the minimum total energy consumption obtained from five different scheduling algorithms: Method 1 (Field "Med. 1"), Method 2 (Field "Med. 2 "), Method 3 (Field "Med. 3"), Method 4 (Field "Med. 4"), and Method 5 (Field "Med. 5"). Column "% M1" and "% M2" represents the percentage of reduction in

3 Types, 3 Modules, T = 400								
Bench.	Num.	Med. 1	Med. 2	Med. 3	Med. 4	Med. 5	% M1	% M2
		E(μJ)	E(μJ)	E(μJ)	E(μJ)	E(μJ)	(%)	(%)
2D(1)	34	1413	1168	1110	1022	925	34.5%	20.8%
2D(2)	4	160	131	124	117	103	35.6%	21.4%
All-pole	29	1090	879	844	781	694	36.3%	21.0%
DPCM	16	606	493	468	432	390	35.6%	20.9%
Floyd	16	685	567	542	508	449	34.5%	20.8%
IIR	16	657	540	516	480	426	35.2%	21.2%
MDFG1	8	353	295	281	260	234	33.7%	20.7%
MDFG2	8	358	298	285	266	235	34.4%	21.1%
WDF(1)	4	169	142	132	121	111	34.3%	21.8%
WDF(2)	12	451	365	353	322	288	36.1%	21.0%
Average Reduction (%)							35.0%	21.1%

TABLE 3.3: The comparison of total energy consumption with five methods while satisfying timing constraint $T = 400$ for various benchmarks.

4 Types, 4 Modules, T = 600								
Bench.	Num.	Med. 1	Med. 2	Med. 3	Med. 4	Med. 5	% M1	% M2
		E(μJ)	E(μJ)	E(μJ)	E(μJ)	E(μJ)	(%)	(%)
2D(1)	34	3091	2518	2436	2256	1963	36.5%	22.1%
2D(2)	4	345	275	271	243	215	37.7%	21.8%
All-pole	29	2354	1902	1842	1687	1481	37.1%	22.0%
DPCM	16	1324	1048	1024	943	817	38.3%	21.9%
Floyd	16	1503	1220	1183	1100	951	36.7%	22.1%
IIR	16	1442	1175	1132	1048	915	36.5%	22.2%
MDFG1	8	778	628	695	564	491	36.9%	21.8%
MDFG2	8	785	631	617	566	493	37.2%	21.9%
WDF(1)	4	372	302	290	278	236	36.6%	21.9%
WDF(2)	12	983	788	764	670	614	37.5%	22.1%
Average Reduction (%)							37.1%	22.0%

TABLE 3.4: The comparison of total energy consumption with five methods while satisfying timing constraint $T = 600$ for various benchmarks.

total energy consumption, compared to Method 1 and 2, respectively. The average reduction is shown in the last row of the table.

The results show that our algorithm LSM can significantly improve the energy reduction on multicore, multi-module architecture. We can see that with more module type selections, the reduction ratio for the total energy consumption has increased. For example, with three types, compared with the VPIS [222], LSM shows an average 21.1% reduction in total energy con-

5 Types, 5 Modules, T = 800								
Bench.	Num.	Med. 1	Med. 2	Med. 3	Med. 4	Med. 5	% M1	% M2
		E(μJ)	E(μJ)	E(μJ)	E(μJ)	E(μJ)	(%)	(%)
2D(1)	34	6159	4998	4827	4395	3791	38.4%	24.1%
2D(2)	4	687	548	536	474	416	39.6%	24.3%
All-pole	29	4705	3806	3651	3329	2896	38.4%	23.9%
DPCM	16	2622	2091	2017	1837	1591	39.3%	23.9%
Floyd	16	2985	2431	2345	2155	1853	37.9%	23.8%
IIR	16	2857	2337	2241	2067	1772	38.0%	24.2%
MDFG1	8	1538	1240	1206	1096	942	38.8%	24.0%
MDFG2	8	1552	1263	1201	1117	961	38.1%	23.9%
WDF(1)	4	736	585	578	516	445	39.5%	23.9%
WDF(2)	12	1960	1558	1537	1362	1179	39.8%	24.3%
Average Reduction (%)							38.8%	24.0%

TABLE 3.5: The comparison of total energy consumption with five methods while satisfying timing constraint $T = 800$ for various benchmarks.

sumption while satisfying timing constraint 200. While using five types, the reduction rate changed to be 24.0% for total energy consumption.

Through the experimental results from Table 3.3 and Table 3.5 we found that VPIS [222] doesn't explore the heterogeneous module type assignment, memory partition, and data prefetching to optimize both energy and performance aspects. Our LSM algorithm combined several novel techniques and can significantly reduce total energy consumption while hiding memory access latency.

In conclusion, LSM algorithm has three main pros: First, we study the combined effects of energy-saving and performance of memory and processing core in a systematic approach. Second, we exploit memory energy saving with loop scheduling and heterogeneous memory module type assignment. Third, performance has been increased compared with the scenario without using our techniques. We use prefetching and keep operations to hide the memory latencies. Memory size usage has been optimized by using memory partition, variable partition, and loop scheduling.

3.5 Conclusion

In this chapter, we present some optimization methods in multicore embedded systems designs. We study the combined effects of energy and performance of memory and processing core in a systematic approach. In these optimization methods, we not only consider optimizing core scheduling, but also partitioning and type assigning for memory in multicore embedded systems designs.

3.6 Glossary

Compiler: A computer program (or set of programs) that transforms source code written in a programming language into the computer machine language.

Data Prefetching: An approach retrieving data from memory and storing it into the cache nearest to the CPU before using it.

Instruction Level Parallelism: A measure of how many of the operations in a computer program can be performed simultaneously.

Locality: The phenomenon of the same value or related storage locations being frequently accessed.

RDRAM: Rambus DRAM, a type of synchronous dynamic RAM, designed by the Rambus Corporation.

Chapter 4

Resource Allocation with Inaccurate Information

4.1 Introduction

The embedded multicore technologies are represented mainly by two categories of multicore processors [118]: 1) processors with dual, quad, and eight cores based on symmetric multiprocessing and 2) processors with combination of heterogeneous cores. An example of the latter kind of multicore is the system on chip (SOC), which has almost unlimited combination of heterogeneous processors on the chip. As the number and the heterogeneity of cores increase, resource allocation management in the embedded multicore system can efficiently improve QoS.

Embedded systems usually operate in environments replete with uncertainties [163]. Meanwhile, these systems are expected to provide a given level of QoS. Stochastic resource allocation can deal with the environment uncertainties and satisfy the QoS demand. In stochastic resource allocation, the uncertainties in system parameters and their impacts on system performance are modeled stochastically. This stochastic model is then used to derive a quantitative evaluation of the robustness of a given resource allocation. This quantitative evaluation results in a probability that the allocation will satisfy the given constraints. A proper approach of stochastic models is using the probability mass functions (PMF) to describe the probability distributions of execution time of tasks running on different cores. According to [10], any claim of robustness for a given system must answer three questions: (a) what behavior of the system makes it robust? (b) What uncertainties is the system robust against? (c) Quantitatively, how robust is the system? For example, some systems are robust if they are capable of finishing all the tasks within a given deadline. A resource allocation deployed in these systems must be robust against uncertainty of the task execution time. The robustness of systems can also be the makespan (total execution time) or the time slackness.

The problem of resource allocation in the field of heterogeneous multicore systems is NP-complete (e.g., [56]). Heuristics are used to find near optimal solutions (e.g., [32, 33, 69, 72, 111, 128, 220]).

In static resource allocations, decisions are made based on the estimated PMFs of execution time of tasks running on different cores. However, when the estimated PMFs of tasks execution time are based on inaccurate information, the estimated PMFs may be different from the actual PMFs. Therefore, the decisions generated by the estimated PMFs may not be robust and the resource allocation is not able to guarantee the given level of QoS. In the first part of this work, a stochastic model for resource allocation is presented. The estimated task execution time information is known as PMF. For a given task schedule, the makespan PMF of a core is generated by convoluting PMFs of all the tasks on its task list. A probability that the whole system can complete all the tasks in a certain time is computed with the makespan PMFs of cores. So for a given resource allocation, we find the robustness, e.g., makespan, that the system can provide with a given probability. We also propose a measurement metric for the impact of the difference between the estimated PMFs and the actual PMFs. In the second part of this work, we stimulate the environment with inaccurate information and compare three greedy heuristics when using the inaccurate information.

In summary, the two major contributions of this work include: 1) the development of a metric for measuring the impact of the inaccurate information on resource allocation; 2) comparing the performance of three heuristics when using incorrect information.

In Section 4.2, we discuss related work. In Section 4.3, models for stochastic task scheduling in multicore embedded systems are presented. We also provide the model for information inaccuracies in this section. We discuss three algorithms for stochastic task scheduling in Section 4.4, followed by experimental results in Section 4.5. Finally, we give the conclusion in Section 4.6.

4.2 Related Work

A framework for robust resource allocation is provided in [10]. In [10], a robustness definition is given. A four-step procedure is established for deriving a robustness metric. In step 1, the robustness of the system is described in a quantitative way, and the range of performance parameter(β_{min}, β_{max}) is given. In step 2, all the system and environmental parameters which may impact the robustness of the system are modeled. In step 3, the relationship between these perturbation parameters and the performance parameters is defined. Finally, the robust range of perturbation parameter is determined by substituting the perturbation parameters in the range of performance parameter(β_{min}, β_{max}).

Previous work has been reported on determining the stochastic behavior of application execution times [59, 121, 26, 160, 167, 168]. In [33], the authors present a derivation of the makespan problem that relies on a stochastic representation of task execution times. In [186], the problem of robust static resource allocation for distributed computing systems under imposed QoS constraints is investigated. A stochastic robustness metric is proposed based on a stochastic model describing the uncertainty in the system and its impact on system performance. Although the stochastic representation of task execution times can describe the system uncertainty, problems arise when modeling the stochastic representation. There are two conventional ways to model the stochastic representation which is usually PMFs: 1) using the statistic information from previous runs of the same tasks to generate the PMFs directly; 2) assuming the PMFs of task execution times are Gaussian distributions, and using the statistic information from previous runs to determine the expectation and the variance [186]. However when the environment is changed, these stochastic representations may not be accurate. For example, a set of PMFs are generated based on some previous runs which occur in a light-weight contention scenario. When they are applied in other heavy contention scenarios, these PMFs are not accurate in the sense that the actual ones may have larger variance due to the heavy contention. So resource allocation with these inaccurate PMFs may lead to the violation of QoS requirement. The previous work above does not evaluate what the relationship is between the degree of inaccuracy in stochastic representation and the degradation of robustness in the system.

4.3 Model and Definition

4.3.1 Stochastic Model

In a normal heterogeneous multicore embedded system, usually there is a set of tasks to be executed. Also, there are a number of cores with various computation power and characteristics in the system. An estimated probabilistic *estimated time to compute* (ETC) matrix P is known before scheduling. For the convenience of the reader, we list the acronyms used in the rest of this chapter in Table 4.1. We assume that the estimated probabilistic ETC matrix is generated using the second approach as discussed in Section 4.2. The entry $P_{i,j}$ of P represents the PMF of execution time of task i on core j. When making the mapping decisions, the information will be used to generate probability distributions of task completion times on different cores.

For a given set of tasks and a given schedule, the *estimated makespan* distribution is the probability distribution of total execution time of the whole set of tasks based on the ETC matrix. We can calculate this probability distribution by convoluting the probability distributions of task execution times. The robustness in this chapter is the minimum makespan (Λ) while maintaining a pre-determined probability θ that all cores will complete their tasks list within Λ.

The estimated PMFs of task execution times are generated with statistic

Name	Description
QoS	Quality of the service
PMF	Probability mass function
ETC	Estimated time to compute
CAT	Core available time
MCT	Minimum completion time alogrithm
M_o	Original makespan
M_n	New makespan
M_c	Correct makespan
MN_o	Normalized original makespan
MN_n	Normalized new makespan
MN_c	Normalized correct makespan
R_n	New_ratio
R_c	Correct_ratio
R_i	Improve_ratio

TABLE 4.1: Acronyms used in the chapter.

information of the previous runs of the same tasks. Any environment or system changes may lead to inaccuracy. Assuming that we can get the updated information about those distributions by some methods, we are able to obtain a resource allocation that meets the QoS requirement with more confidence. We call these distributions (PMFs) the updated PMFs. There are methods to obtain the updated PMFs, for example, establishing the quantitative relationship between the change of environment parameters and the change of distribution of task execution times. The development of these methods is out of the scope of this chapter.

In the case that we can get the updated PMFs of task execution times, whether a new resource allocation is necessary becomes another problem. Using a new resource allocation not only requires time to re-run the scheduling algorithm, but also brings the overhead of rearranging resources in the system. However, if we can predict the degradation of robustness based on the difference between the updated PMFs and the estimated PMFs, i.e., the degree of inaccurate information, we can decide whether a new resource allocation is necessary. Furthermore, with knowledge of which scheduling algorithm performs the best when using inaccurate information, we can reduce the probability that a new resource allocation is necessary by using the best scheduling algorithm. We will provide some insight on these two questions in our evaluation part in the chapter.

4.3.2 Measurement Parameters

Since the difference between the estimated PMFs and the updated PMFs may cause the robustness degradation, several measurement parameters are introduced to measure the robustness degradation.

- *Original Schedule*: Task schedule generated by using the estimated PMFs
- *Remapped Schedule*: Task schedule generated by using the updated PMFs
- *Makespan*: The total time taken for a system to finish all the tasks with a given task schedule
- *Original Makespan (M_o)*: The makespan using the estimated PMFs and the original schedule
- *New Makespan (M_n)*: The makespan using the updated PMFs and the original schedule
- *Correct Makespan (M_c)*: The makespan using the updated PMFs and the remapped schedule
- *New_ratio (R_n)*:

$$R_n = \frac{M_n - M_o}{M_o}$$

- *Correct_ratio (R_c)*:

$$R_c = \frac{M_c - M_o}{M_o}$$

- *Improve_ratio (R_i)*:

$$R_i = \frac{M_n - M_c}{M_c}$$

As discussed in the previous section, the robustness metric in this chapter is the minimum makespan (Λ) while maintaining a predetermined probability θ that all cores will complete their tasks list within Λ. The smaller the makespan (Λ) is, the more robust the system is. Original makespan gives the robustness of the system assuming accurate information is used in the scheduling. When inaccurate information is used in the original scheduling, new makespan results in the actual robustness of the system without rerunning the scheduling algorithm. Correct makespan indicates the new robustness when a new schedule is generated with updated accurate information. New_ratio shows the degradation of the robustness when using the inaccurate information. Improve_ratio reveals the improvement caused by rerunning the scheduling algorithm. Correct_ratio indicates the impacts of the changes of environment on the system's robustness.

4.4 Algorithms

4.4.1 Overview

Three static greedy heuristics are used. *Minimum completion time* (MCT) [14] is a one-phase heuristic. The output of this heuristic depends on the order in which the tasks are mapped to the cores. *Min-min* [14, 74] and *max-min* [14, 74] are two-phase heuristics. These two heuristics are independent from the tasks assigning order in the sense that for a given set of tasks and a system with a certain set of cores, the outputs are identical no matter how many times it runs.

Greedy heuristics are widely used in heterogeneous system resource allocation. Compared to global heuristics such as *genetic algorithm* and *simulated annealing*, greedy heuristics can get a schedule much quicker than global heuristics. Previous work shows that min-min heuristics can get a schedule as optimal as the one generated by a genetic algorithm.

The definitions of the three heuristics are provided below. *Core available time* (CAT) is the probability distribution of time when the core will finish all the tasks which are assigned to this core previously. The PMF of the completion time for a new task t_i on core c_j, $ct_{i,j}$, can be calculated by convoluting the CAT of core c_j and the execution time distribution of task t_i on core c_j.

4.4.2 MCT

Minimum Completion Time (MCT) [14] assigns tasks in an arbitrary order to the cores. For an unmapped task, MCT maps it on the core which can complete this task in the earliest time while maintaining a certain probability. The idea behind MCT is that it considers both the execution time of the task on the core as well as load balance. Since MCT assigns tasks in an arbitrary order, the scheduling results are non-determinstic. We present the MCT algorithm in Figure 4.1.

4.4.3 Min-Min

Min-min [14, 74] selects the task-core pair in two phases. In phase 1, for each unmapped task, the core which can complete it in the earliest time while maintaining a certain probability is selected to form a pair. In phase 2, among all the pairs, the pair that has the minimum *ct* is selected, and the task in the pair is mapped to the according core. The idea behind min-min is that it does its best to keep the current load balance with the least change on it. The min-min is provided in Figure 4.2.

Require: A set of tasks, m different cores, ETC PMF matrix
Ensure: A MCT resource allocation schedule
1: A list of unmapped tasks U is generated.
2: Reorder the list in an arbitrary order.
3: **while** the list U is not empty **do**
4: The first task i in the list U is selected, then among the m cores, the core j which has the minimum $ct_{i,j}$ is also selected.
5: Assign the task to the core.
6: Remove the task from the list U.
7: Update the CAT of the selected core.
8: **end while**

FIGURE 4.1: MCT algorithm.

Require: A set of tasks, m different cores, ETC PMF matrix
Ensure: A min-min resource allocation schedule
1: A list of unmapped tasks U is generated.
2: **while** the list U is not empty **do**
3: For each task in the list U, find the core that gives the minimum ct.
4: Among task-core pairs formed in step 3, find the pair with the minimum ct.
5: Assign the task in the selected pair to the according core.
6: Remove the task from the list U.
7: Update the CAT of the selected core.
8: **end while**

FIGURE 4.2: Min-min algorithm.

4.4.4 Max-Min

Max-min [14, 74] is similar to min-min. In phase 1, max-min does exactly the same as min-min. Then in phase 2, max-min finds the task-core pairs with the maximum ct, which is different from min-min. The idea behind this is that tasks with larger execution time will likely increase the penalty if these tasks are not assigned to their best cores. Figure 4.3 shows the max-min algorithm.

Require: A set of tasks, m different cores, ETC PMF matrix
Ensure: A max-min resource allocation schedule
1: A list of unmapped tasks U is generated.
2: **while** the list U is not empty **do**
3: For each task in the list U, find the core that gives the minimum ct.
4: Among task-core pairs formed in step 3, find the pair with the maximum ct.
5: Assign the task in the selected pair to the according core.
6: Remove the task from the list U.
7: Update the CAT of the selected core.
8: **end while**

FIGURE 4.3: Max-min algorithm.

4.5 Simulation

4.5.1 Simulation Setup

To evaluate the robustness degradation caused by the inaccurate information, the following approach was used to simulate the stochastic resource allocation in a heterogeneous multicore embedded system. A set of 1024 independent tasks was formed randomly. They consist of 28 task classes, where tasks in the same class are identical. There are eight heterogeneous cores in a system. Each of these cores has its own computation power and characteristic. So the estimated probabilistic ETC matrix P has the size of 28×8. PMF $P_{i,j}$ is based on gamma distribution with a mean of $m_{i,j}$ and a standard deviation of $sd_{i,j}$. In our simulation, we generate PMFs by sampling the probability density functions (PDF) of gamma distributions with a start point, an end point and a fixed step. Each of the 40 simulation trials has different estimated probabilistic ETC matrix P.

Before sampling the PDF of gamma distributions, the value of mean and standard deviation needs to be determined. We randomly generate a 28×8 mean matrix based on gamma distribution as well as the standard deviation matrix. Here, we use the COV based method [11] with the mean of task execution time from 40 to 80, and both coefficients of variation of tasks and cores uniformly from 0.35 to 1. When forming the PMF $P_{i,j}$, we can sample the PDF of gamma distribution with a mean of $m_{i,j}$ and a standard deviation of $sd_{i,j}$.

To simulate the case in which updated PMFs are different from the estimated PMFs, parameters (mean or standard deviation) of the updated PMFs are generated by multiplying the parameters of the estimated PMFs with a scalar matrix S.

For example, if the means are modified,

$$updated\_mean(i,j) = mean(i,j) \times S_{i,j}$$

The entry of scalar matrix S is based on a uniform distribution with a range of $[S_{min}, S_{max}]$.

4.5.2 Simulation Results

4.5.2.1 Compare the Impacts on Robustness When Modifying Different Parameters

In this part, we compare the impacts on robustness when using different scalar matrixes as well as modifying different parameters.

We simulate two different scenarios in which two different kinds of inaccurate information occur:

1. Keep the standard deviations unchanged, and multiply the means with a scalar matrix.

2. Keep the means unchanged, and multiply the standard deviation with a scalar.

The first scenario usually happens when the embedded system is employed in a physically inconstant environment. For example, in an environment where temperature changes rapidly, cores will likely run faster in low temperature than in high temperature. As the temperature increases, the means of the probability distribution of execution times may increase. In this case, the statistic information collected previously in low temperature may not be accurate. The second scenario happens when resource contention among tasks changes. When the resource contention is light, a core likely finishes the same tasks in a narrow distribution, especially around the mean of distribution. When the contention is heavy, the distribution of a task class in a core may be wide, i.e., with larger standard deviations. In our simulation, the scalar matrixes are with ranges of [0.1, 1.9], [0.1, 2.9], [0.1, 3.9], [0.1, 4.9].

MCT heuristic is used in all these four parameter modifications. The result of each trial is the average value of MCT with 25 different task mapping orders.

In Figure 4.4, the increase of new_ratio is proportional to the increase of the scalar matrix range with 20% to 70% penalty. Obviously, the increase of mean values of execution time distribution leads to a longer makespan. This 20% to 70% penalty is caused by the inaccurate information used in the original schedule. We find that the improve_ratio, which indicates the improvement of rescheduling, does not change as much as the increase of the scalar matrix range. Note that when we calculate the improve_ratio, we compare the difference between the new_makespan and the correct_makespan. In the

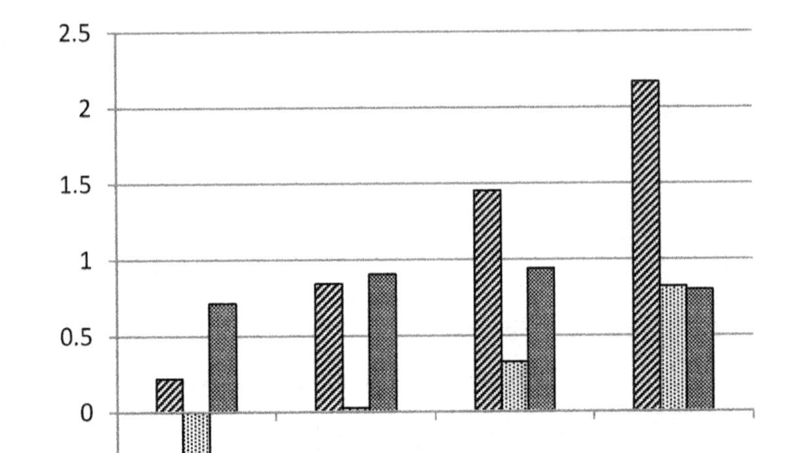

FIGURE 4.4: New_ratio, correct_ratio and improve_ratio when changing the mean.

convolution of these two distributions, we use the updated PMFs. What the "Improve_ratio" columns show us is that how much the rescheduling improves does not mainly depend on how inaccurate the information is, but depends on what the task set consists of. The correct_ratio is also proportional to the increase of the scalar matrix range. It shows that the degradation of robustness is a linear function of the degree of how the environment changes. Comparing Figure 4.5 to Figure 4.4, we find that the inaccurate standard deviations have much less impacts on the robustness than the inaccurate means do.

4.5.2.2 Compare the Performance of Different Heuristics

In this part, three different heuristics (min-min, MCT, max-min) are compared with their performance when using inaccurate information. In this part, we will keep the standard deviations fixed and change the means. To compare the performance of these heuristics, normalized makespan of MCT and max-min is introduced.

- Max-min normalized original makespan

$$MN_o(Max - min) = \frac{M_o(Max - min)}{M_o(Min - min)}$$

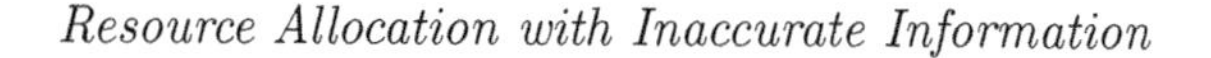

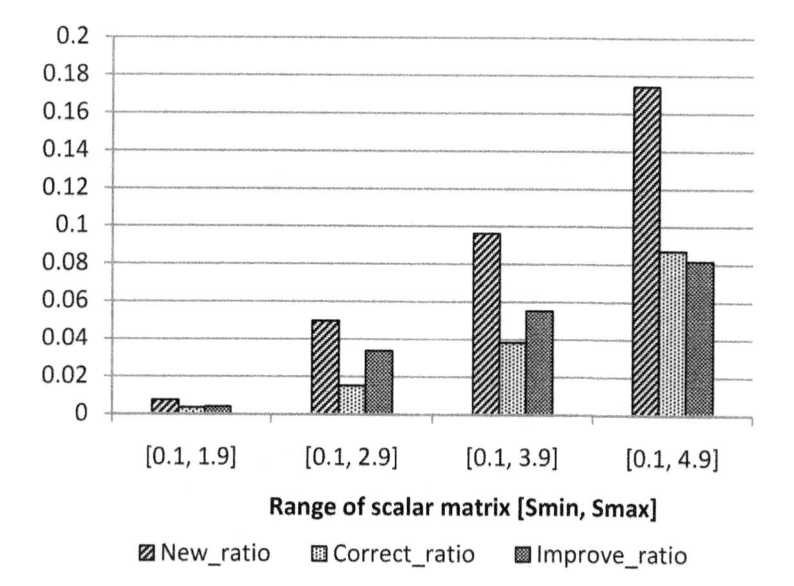

FIGURE 4.5: New_ratio, correct_ratio and improve_ratio when changing the standard deviation.

- Max-min normalized new makespan

$$MN_n(Max - min) = \frac{M_n(Max - min)}{M_n(Min - min)}$$

- Max-min normalized correct makespan

$$MN_c(Max - min) = \frac{M_c(Max - min)}{M_c(Min - min)}$$

- MCT normalized original makespan

$$MN_o(MCT) = \frac{M_o(MCT)}{M_o(Min - min)}$$

- MCT normalized new makespan

$$MN_n(MCT) = \frac{M_n(MCT)}{M_n(Min - min)}$$

- MCT normalized correct makespan

$$MN_c(MCT) = \frac{M_c(MCT)}{M_c(Min - min)}$$

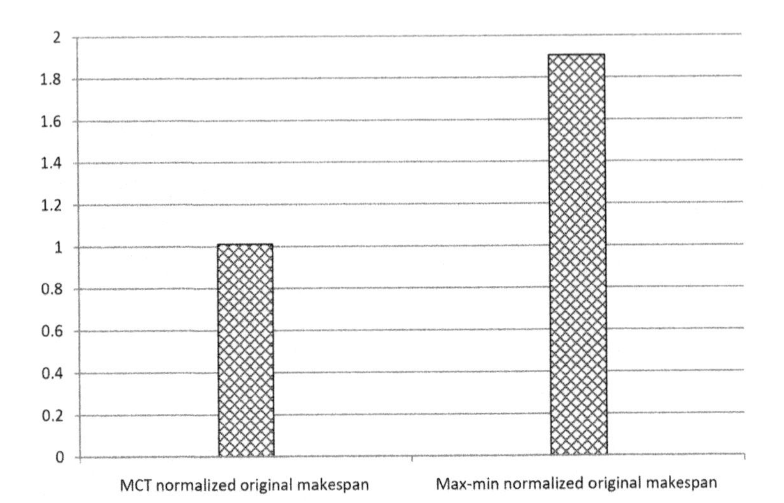

FIGURE 4.6: Original makespans comparison.

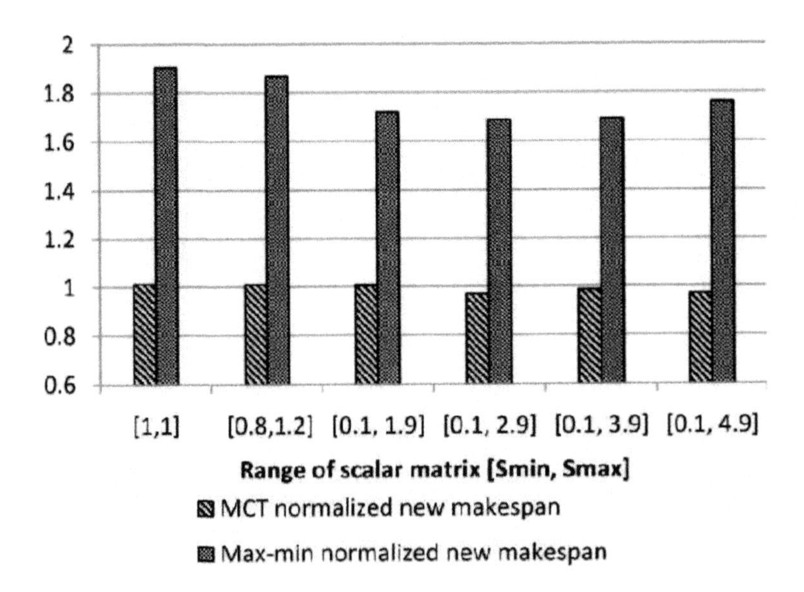

FIGURE 4.7: Normalized new makespans comparison.

In Figures 4.6 to 4.8, max-min has the longest makespan among these three heuristics. So the max-min performance is the worst among these three heuristics. The performance of MCT is very close to the performance of min-min with respect to the original makespan. Furthermore, MCT outperforms the min-min in the new makespan. It means that MCT is less impacted by the inaccurate information and performs close to the min-min in the original makespan, and it performs the best in the new makespan even though the difference between these two heuristics is not significant.

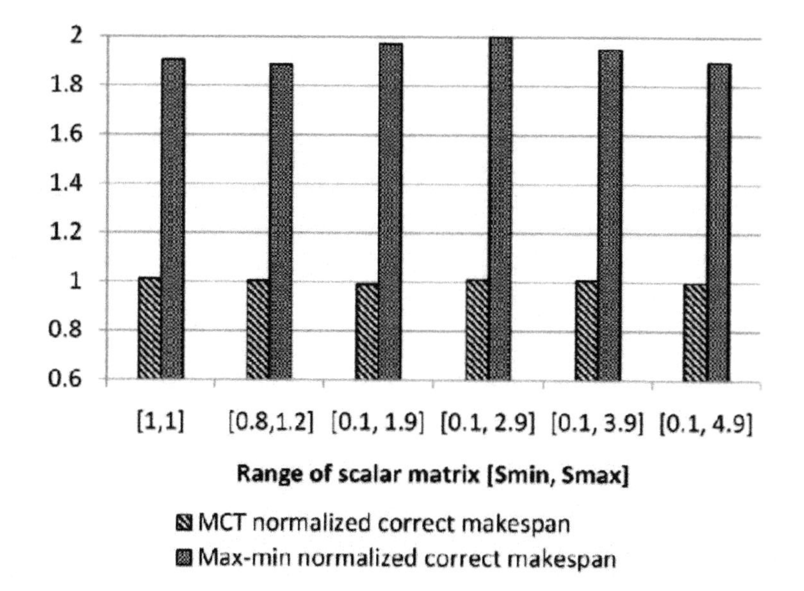

FIGURE 4.8: Normalized correct makespans comparison.

4.6 Conclusion

We have proposed a systematic measure of the robustness degradation in a stochastic way. We have evaluated the impacts of inaccurate information on system robustness in two different scenarios. In our simulation, the makespan is the robustness metric. We find that the makespan with inaccurate information increases proportional to the increase of the means of task execution time distribution caused by environment changes. Also, 20% to 70% penalty is caused by the inaccurate information used in making scheduling decisions. The impact of environment changes on the robustness is linear to the degree of how much inaccurate information (mainly the shift of means of the PMFs) is generated by these environment changes. However the improvement of re-scheduling with updated information mainly depends on what the task set consists of, not how inaccurate the information is. We also find that the impact of inaccurate means of the PMFs is much larger than inaccurate standard deviations.

Among these three greedy algorithms, MCT performs the best under inaccurate information. It generates schedules which are almost as optimal as the ones from min-min where accurate information is used. And inaccurate information has less impact on schedules from MCT than it does on min-min. max-min performs the worst.

4.7 Glossary

Gamma Distribution: A well-known two-parameter family of continuous probability distributions.

Greedy Heuristics: A greedy heuristic is any algorithm that follows the problem solving metaheuristic of making the locally optimal choice at each stage [1] with the hope of finding the global optimum.

Makespan: The total time taken for a system to finish all the tasks with a given task schedule.

PMF: A probability mass function is a function that gives the probability that a discrete random variable is exactly equal to some value.

QoS: Quality of the services. It refers to the ability to provide different priority to different applications, users, or data flows, or to guarantee a certain level of performance to a data flow.

SOC: System on chip refers to integrating all components of a computer or other electronic system into a single integrated circuit (chip).

Chapter 5

Heterogeneous Parallel Computing

5.1 Introduction

Embedded systems, which has tight conjoining of and coordination between computational and hardware resources, is usually heterogeneous. This chapter focuses on the optimization for soft real-time heterogeneous embedded systems. Hardware technology provides the potential of designing flexible platforms with high performance or low power, but with the current software technology, which does not take advantage of hardware platform, it will be hard to obtain an acceptable solution. The boundary between software and hardware parts is more obscured than ever. Conducting research in system optimizations while considering hardware is an important step to advance the development of heterogeneous embedded systems.

The design of heterogeneous embedded systems poses a host of technical challenges different from those faced by traditional computation and communication systems because heterogeneous embedded systems have much larger design space and are more complicated to optimize in terms of cost, timing, and power [87, 55, 158]. The optimization of realistic embedded systems requires efficient algorithms and tools that can handle multiple objectives simultaneously. Since the performance of embedded systems depends on the appropriate configuration and efficient use of hardware [127], [66], [105], we need to consider both software components and hardware components (platforms). In this chapter, we will design new techniques that consider timing and cost optimization on software/hardware collectively for distributed and heterogeneous embedded systems.

With the rapid advance of the hardware technology — easy configuration of different types of hardware and system on a chip — nowadays we can put multiple cores and configure different types of functional units and memories in one chip. For example, Cell processor has one POWER processor dedicated to the operating system and other control functions, and eight synergistic cores used for computations [86]. Other examples include network processors, DSP processors as well as networked sensor networks and many portable devices.

Today, embedded systems are becoming increasingly complex. Not only is there increased functionality, but the system architecture is also becoming more complex. The design task now is dominated by *system integration*, namely composing a set of high level components and some application specific components [55]. Since hardware-manufacturing cycles do take time and are expensive, the interest in software-based implementation has risen to previously unseen levels. The increase in computational power of processors and the corresponding decrease in size and cost have allowed moving more and more functionality to software [215].

Components are pre-implemented software modules, which are treated as building blocks in integration. The integrated embedded system can be viewed as a collection of communicating reusable components. Figure 5.1 shows the embedded software constructed by integrating components. Figure 5.1(a) shows the basic structure of embedded systems. Therefore, different platforms will greatly affect the software design. Figure 5.1(b) shows the embedded software construction [221].

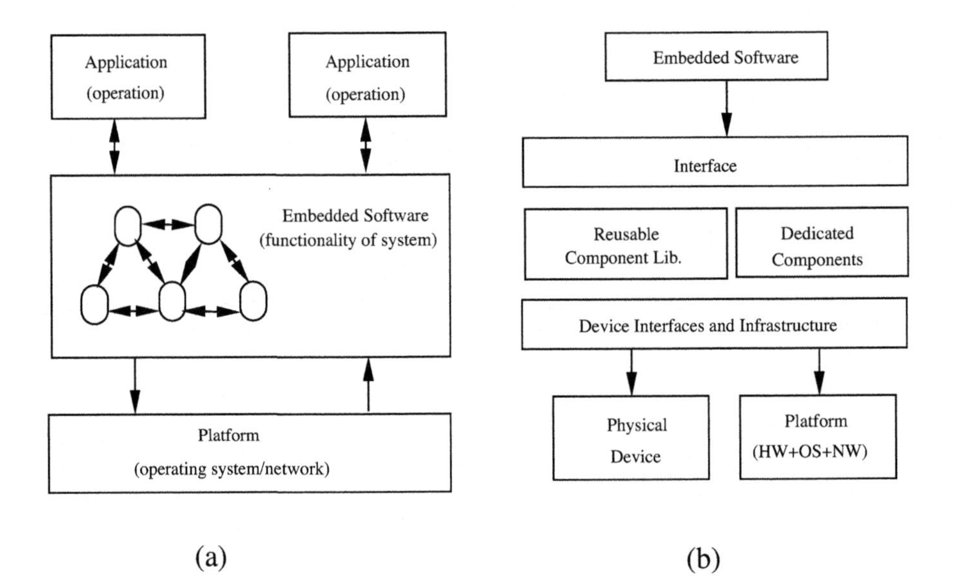

FIGURE 5.1: (a) Embedded system structure. (b) Embedded software construction.

For heterogeneous parallel embedded systems, two scenarios need to be studied: fixed hardware platform and configurable hardware platform. There are more and more software components and hardware platforms (components) available to be selected for using [221], [215], [201], [78]. For a fixed hardware platform, it is critical to select a proper component for each task for real-time embedded systems so that the cost can be minimized while the timing constraint can be satisfied. This problem has been studied extensively. But with the advance of hardware technology, we can configure and assemble hardware components as well as software components.

The field of *component-based software engineering* (CBSE) is being looked to as the emerging paradigm that will bring rigor to the software development process [209], [197], [228], [47], [34], [216], [157]. But little research has been conducted to consider the configurability of hardware platforms while doing optimizations. This is a great loss without considering hardware platforms selection. The interoperability of the components has been studied extensively on CBSE. This chapter will not focus on interoperability but on performance as the research in [9], [36] did. We will choose both hardware and software components simultaneously; and study how to minimize cost while satisfying timing constraints by selecting best configuration of software/hardware components (platforms).

For real-time applications, execution time is a major concern. We need to carefully consider how to realistically model execution times. Tasks may have variable execution times due to conditional instructions and/or operations that could have different execution times for different inputs [207], [247], [91] in software/hardware components. Existing methods are not able to deal with such uncertainty effectively. Therefore, either worst-case or average-case computation times for these tasks are usually assumed. Such assumptions, however, may result in an inferior task assignment. It is not easy to deal with uncertainty in an optimization process while the performance of results needs to be guaranteed. In our chapter [160], we adopt a probabilistic approach, but only consider either configurable software or hardware platforms. Comparing with previous work, our current approach is more complicated and complete. This chapter will model the uncertainty by probabilistic model and design new optimization algorithms that can guarantee the performance of results in a systematic approach.

Data flow graph (DFG) model is commonly used to model DSP and computation sequences. Each node in DFG represents a task with execution time. Edges in DFG represent dependencies between tasks. To model a realistic situation, the execution time of a node executed by a cyber/physical component should not be considered fixed. In this chapter, we will model each varied execution time as a probabilistic random variable and define the component selection problem as *heterogeneous assignment with probability* (HAP) problem. That is, we will select hardware platform (component) and software components simultaneously. The solution of the HAP problem assigns a proper hardware platform and software component to each task such that the total cost is minimized while the timing constraint is satisfied with a guaranteed confidence probability. This study will be useful for soft real-time embedded systems. If we don't consider probabilistic times, we may get inferior designs and miss many good design opportunities.

Heterogeneity is one of the most important features of modern embedded systems, which become more and more complicated. In many embedded systems, such as heterogeneous parallel DSP systems [88], [19], the same type of operations can be processed by heterogeneous software/hardware components with different costs. These costs may relate to power, reliability, etc. Therefore, an important problem arises: how to assign a proper component type to each operation of a software application such that the requirements can be met and the total cost can be minimized while satisfying timing constraints with a guaranteed confidence probability.

We call the assignment problem that uses the model of fixed execution time as the hard HA (*hard Heterogeneous Assignment*) problem and the one that uses the model of non-fixed execution time as the HAP (*Heterogeneous Assignment with Probability*) problem. Using probabilistic approach, we can obtain solutions that can be used for hard real-time systems as well as providing multiple superior choices for soft real-time systems while timing constraints are satisfied with guaranteed confidence probabilities. It is easy to see that this problem is a NP-complete problem.

We model the embedded systems as two types according to hardware platforms. If the hardware platform is fixed, we call it the *fixed platform model.* Otherwise, we call it the *configurable platform model.* The execution time of each node maybe fixed or non-fixed. If the execution time is fixed, we call it the *fixed execution time model.* If the execution time of each node has probability distribution, we model the execution time of each task as a random variable and call this type the *non-fixed execution time model.* So there are four combinations: 1) fixed platform, fixed execution time model, 2) fixed platform, non-fixed execution time model, 3) configurable platform, fixed execution time model, 4) configurable platform, non-fixed execution time model.

Since the performance of the embedded systems depends on the appropriate configuration and efficient use of different components of the whole system, we need to consider both computation and communication of the embedded systems. In this chapter, we have considered both the energy cost and delay of each communication.

Our contributions to this chapter are listed as the following:

- We design embedded systems with the consideration of configurable hardware platforms.
- We exploit the soft real-time and non-fixed execution time to improve energy-saving.
- We propose the OESCP algorithm with these two considerations to provide an overall cost optimization without sacrificing performance.
- We consider both energy cost and delay of each communication in our models.

The rest of this chapter is organized as following: In Section 5.2, we give a motivational example. The models and basic concepts are introduced in Section 5.3. In Section 5.4, we propose the OESCP algorithm. The experimental results are shown in Section 5.5.

5.2 Motivational Example

5.2.1 Fixed Hardware Platform Model of Heterogeneous Embedded Systems

In the HAP problem, we model the execution time of a task as a random variable [160]. For heterogeneous embedded systems, each component is associated with a cost related to hardware cost, area, reliability, power consumption, etc. Faster one has higher cost while slower one has lower cost. We need to study how to assign a proper software/hardware component to each node of a DFG such that the total cost is minimized while the timing constraint is satisfied with a guaranteed confidence probability. With a confidence probability P, we can guarantee that the total execution time of the DFG is less than or equal to the timing constraint with a probability greater than or equal to P.

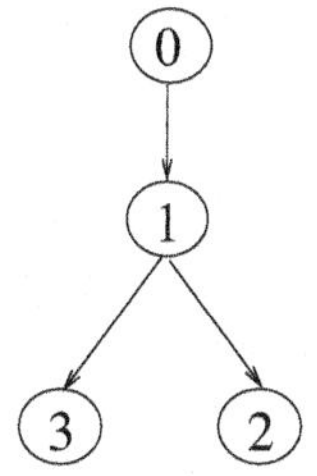

(a)

Nodes	R1			R2		
	T	P	C	T	P	C
0	1	1.0	9	2	1.0	5
1	1	0.9	10	2	0.7	4
	3	0.1	10	4	0.3	4
2	1	0.9	9	2	0.8	5
	4	0.1	9	6	0.2	5
3	1	0.1	8	3	0.3	3
	2	0.9	8	6	0.7	3

(b)

Nodes	R1			R2		
	T	P	C	T	P	C
0	1	1.0	9	2	1.0	5
1	1	0.9	10	2	0.7	4
	3	1.0	10	4	1.0	4
2	1	0.9	9	2	0.8	5
	4	1.0	9	6	1.0	5
3	1	0.1	8	3	0.3	3
	2	1.0	8	6	1.0	3

(c)

FIGURE 5.2: (a) A given tree. (b) The time, probabilities, and costs of its nodes for different component types. (c) The time cumulative distribution functions (CDFs) and costs of its node for different component types.

We show an example to illustrate the HAP problem. Assume that we can select from two hardware platforms for each node: R_1, R_2. An exemplary DFG is shown in Figure 5.2(a), which is a tree with four nodes. The execution time (T), corresponding probabilities (P), and costs (C) of each node for different software components are shown in Figure 5.2(b). Each node can choose one of two components, and executes with probabilistic execution times. Thus, execution time (T) of each component is a random variable. For example, when choosing R_1, node 1 can be executed in 1 time unit with probability 0.9 and be executed in 3 time units with probability of 0.1. In other words, node 1 can be finished in 3 time units with 100% probability. Figure 5.2(c) shows the time *cumulative distribution functions* (CDFs) and costs of each node for different component types. For demonstration purposes, in this example, we can add a fixed number for the energy cost and delay of each communication; assuming both are 1, the conclusion will not be changed.

T	(P , C)	(P , C)	(P , C)	(P , C)	(P , C)	(P , C)
3	0.08, 36					
4	0.06, 30	0.72, 32	0.81, 36			
5	0.56, 26	0.72, 28	0.81, 32			
6	0.17, 21	0.56, 22	0.63, 26	0.72, 28	0.81, 32	0.90, 36
7	0.17, 17	0.19, 21	0.56, 22	0.80, 26	0.90, 30	
8	0.17, 17	0.24, 21	0.80, 22	0.90, 26	**1.00, 36**	
9	0.24, 17	0.70, 21	0.80, 22	0.90, 23	**1.00, 30**	
10	0.70, 17	0.80, 22	0.90, 23	**1.00, 26**		
11	0.70, 17	**1.00, 21**				
12	**1.00, 17**					

TABLE 5.1: Minimum total costs with computed confidence probabilities under various timing constraints for a tree.

The minimum total costs with computed confidence probabilities under various timing constraints for a tree are shown in Table 5.1. In Figure 5.2, if we use the worst-case execution time as the fixed execution time for each node, then the assignment problem becomes the *hard heterogeneous assignment* (hard HA) problem, which is related to the hard real-time system. With certain timing constraints, there might be no solution for the worst-case situation. In Table 5.1, the entries with probability equal to 1 (see the entries in boldface) actually give the results to the hard HA problem which shows the worst-case scenario of the HAP problem. However, for soft real-time applications, it is desirable to find an assignment that guarantees the total execution time can be finished within the timing constraint with a provable probability.

For example, in Table 5.1, with timing constraint 7, we cannot find a solution to the hard HA problem. But we can find an assignment satisfying the timing constraint 7 with probability of 0.9. Also, the cost can be reduced from the assignment by using our probabilistic approach. For example, at timing constraint 8, the minimum cost is 36 for the hard HA problem. while solving the HAP problem, with 90% confidence probability of satisfying timing constraint 8, we get the minimum cost of 26, which gives 27.8% improvement. The assignments for each pair of (0.90, 26) and (1.00, 36) under the timing constraint 8 are shown in Table 5.2.

5.2.2 Configurable Hardware Platform, Fixed/Non-Fixed Execution Time Model

In many embedded systems, there are several hardware platforms available, that is, the hardware platforms are configurable. For each hardware platform, there are several software components available. When we co-select a certain

	Node id	Type id	T	Probability	Cost
(0.90, 26)	0	2	2	1.00	5
	1	2	4	1.00	4
	2	1	1	0.90	9
	3	1	2	1.00	8
Total			8	0.90	26
(1.00, 36)	0	1	1	1.00	9
	1	1	3	1.00	10
	2	1	4	1.00	9
	3	1	2	1.00	8
Total			8	1.00	36

TABLE 5.2: With timing constraint 8, the assignments of types for each node with different (Probability, Cost) pairs.

hardware platform and software component, the execution time may or may not be fixed. We will give an example which shows the configurable hardware platform non-fix execution time model.

1 → 2

(a)

Nodes	H1						H2					
	S1			S2			S1			S2		
	T	P	C	T	P	C	T	P	C	T	P	C
1	1	0.9	10	2	0.7	4	1	0.9	15	1	0.7	8
	3	0.1	10	4	0.3	4	2	0.1	15	3	0.3	8
2	1	0.9	9	2	0.8	5	1	0.9	12	2	0.7	10
	4	0.1	9	6	0.2	5	3	0.1	12	4	0.3	10

(b)

FIGURE 5.3: (a) A given DFG. (b) The time, probabilities, and costs of its node for different hardware platforms and software components.

An exemplary DFG is shown in Figure 5.3(a), which is a simple path with two nodes. Assume that we can select from two hardware platforms (H_1, H_2) and two software components (S_1, S_2) for each node. The execution time (T), probabilities (P), and costs (C) of each node for different hardware platforms and software components are shown in Figure 5.3(b). Each node can choose one of two platforms and one of two components, and executes with probabilistic execution times. Thus, execution time (T) of each component is a random variable. For example, when choosing platform H_1 and component S_1, node 1 can be executed in 1 time unit with probability 0.9 and be executed in 3 time units with probability of 0.1. In other words, node 1 can be finished in 3 time units with 100% probability.

In Figure 5.3, under different models, there are different optimization results. It is very important to exploit the advantages of configurable platform models. For example, with timing constraint 5, we have: 1) For configurable platform, non-fixed execution time model, we can obtain minimum cost 13 with 90% guaranteed probability. The assignment is: node 1 (H_1 and S_2 with $T = 4$); node 2 (H_1 and S_1 with $T = 1$). 2) For configurable platform, fixed execution time model, the minimum cost is 27. The assignment for both nodes is H_2 and S_1. 3) For fixed platform, non-fixed execution time model, assume node 1 selects platform H_2 and node 2 selects H_1, we can obtain minimum cost 17 with 90% guaranteed probability. The assignment is: node 1 (S_2 with $T = 3$); node 2 (S_1 with $T = 1$). 4) For fixed platform, fixed execution time model, assume node 1 selects platform H_2 and node 2 selects H_1, there is no solution. Comparing 1) with 2) there is 51.9% cost saving. Comparing 1) with 3) there is 23.5% cost saving. The detail assignments of each node for all four models under the timing constraint 5 are shown in Table 5.3. Column "N_id", "H_id", and "S_id" represent "Node id", "Hardware id", and "Software id", respectively.

	N_id	H_id	S_id	T	Probability	Cost	Saving
Model 4	1	1	2	4	1.00	4	
	2	1	1	1	0.90	9	-
Total				5	0.90	13	
	1	2	1	2	1.00	15	
Model 3	2	2	1	3	1.00	12	51.9%
Total				5	1.00	27	
	1	2	1	3	1.00	8	
Model 2	2	1	1	1	0.90	9	23.5 %
Total				5	0.90	17	
	1	2					
Model 1	2	1	no solution				-
Total							

TABLE 5.3: With timing constraint 5, the component-assignments of each node for four models.

5.3 System Model

Prior design space exploration methods for hardware/software codesign of embedded systems guarantee no deadline missing by considering worst-case execution time of each task [218], [97], [98]. These methods are pessimistic and are suitable for developing systems in a hard real-time environment, where any deadline miss will be catastrophic. However, they cannot effectively apply to soft real-time systems, such as heterogeneous video systems, which can tolerate occasional violations of timing constraints. For example, in packet audio applications, loss rates between 1% to 10% can be tolerated [30]. They often lead to over-designed systems that deliver higher performance than necessary at the cost of expensive hardware, high energy consumption, and other system resources.

Modeling execution time in probabilistic distribution has been studied for soft real-time systems design [162, 247, 100, 205, 92, 91, 90, 207]. The general assumption is that the execution time of a selected component for each task can be described by a discrete probability density function that can be obtained by applying path analysis and system utilization analysis techniques. We can obtain the probability distribution of execution time for each task by sampling and knowing detailed timing information about the system or by profiling the components [205], [159].

Incorporating probability into optimization is not easy. Hu and Sha et al. [247] proposed a state-based probability metric to evaluate the overall probabilistic timing performance of the entire task set. However, their evaluation method becomes time consuming when tasks have many different execution time variations. Hua et al. [92], [91] proposed the concept of *probabilistic design* where they design the system to meet the timing constraints of periodic applications statistically. But their algorithm is not optimal and only suitable for uniprocessor executing tasks according to a fixed order, that is, a simple path. *Probabilistic data flow graph* (PDFG) was studied by Tongsima, Sha et al., [207], [153]. PDFG is defined as a DFG where execution time of each node is represented by a probabilistic random variable. Given a PDFG, each node only has one random variable representing computation time, so they do not consider component selection (assignment).

In the HAP problem, we model the execution time of a component associated with a node as a random variable. There might be multiple components that can be assigned for each node. And each assignment is associated with a random variable. We call this graph model as *Heterogeneous Probabilistic Data Flow Graph* (HPDFG). For example, a node can be assigned by two different components. Each assignment is associated with a random distribution of execution time. A ***HPDFG G*** $= \langle V, E\rangle$ is a *directed acyclic graph* (DAG), where $V = \langle v_1, v_2, \cdots, v_N\rangle$ is the set of nodes, $E \subseteq V \times V$ is the edge set that defines the precedence relations among nodes in V. In practice, many architectures consist of different types of components. Assume there are maximum M different components in a components set R=$\{R_1, R_2, \cdots, R_M\}$. For each component, there are maximum K execution time variations T, although each node may have a different number of components and execution time variations.

An assignment for an HPDFG G is to assign a component type to each node. Define an ***assignment A*** to be a function from domain V to range R, where V is the node set and R is component type set. For a node $v \in V$,

$A(v)$ gives the selected type of node v. For example, in Figure 5.2, assigning component types 2, 2, 1, and 1 for nodes 0, 1, 2, and 3, respectively, we obtain minimum total cost 26 with 0.9 probability satisfying the timing constraint 8. That is, $A(0) = 2$, $A(1) = 2$, $A(2) = 1$, and $A(3) = 1$.

In a HPDFG G, each varied execution time T is modeled as a probabilistic random variable. $\mathbf{T_{R_j}(v)}$ $(1 \leq j \leq M)$ represents the execution times of each node $v \in V$ for component type j, and $\mathbf{P_{R_j}(v)}$ $(1 \leq j \leq M)$ represents the corresponding probability function. And $\mathbf{C_{R_j}(v)}$ $(1 \leq j \leq M)$ is used to represent the cost of each node $v \in V$ for component type j, which is a fixed value. For instance, in Figure 5.2, $T_1(1) = 1, 3$; $T_2(1) = 2, 4$. Correspondingly, $P_1(1) = 0.9, 0.1$; $P_2(1) = 0.7, 0.3$. And $C_1(1) = 10$; $C_2(1) = 4$.

Given an assignment A of a HPDFG G, we define the ***system total cost under assignment A***, denoted as $\mathbf{C_A(G)}$, to be the summation of costs of all nodes, that is, $C_A(G) = \sum_{v \in V} C_{A(v)}(v)$. In this chapter we call $C_A(G)$ as ***total cost*** in brief. For example, in Figure 5.2, under assignment 2, 2, 1, and 1 for nodes 0, 1, 2, and 3, respectively, the costs of nodes 0, 1, 2, and 3 are: $C_2(0) = 5$, $C_2(1) = 4$, $C_1(2) = 9$, and $C_1(3) = 8$. Hence, the total cost of the graph G is: $C_A(G) = C_2(0) + C_2(1) + C_1(2) + C_1(3)$, that is, $C_A(G) = 26$.

For the input HPDFG G, given an assignment A, assume that $\mathbf{T_A(G)}$ stands for the ***execution time of graph G under assignment A***. $T_A(G)$ can be gotten from the longest path p in G. The new variable $T_A(G) = \max_{\forall p} T_{A(v)}(p)$ is also a random variable, where $T_{A(v)}(p) = \sum_{v \in p} T_{A(v)}(v)$. In Figure 5.2, there is only one path. Under assignment 2, 2, 1, and 1 for nodes 0, 1, 2, and 3, $T_A(G) = T_{A(v)}(p) = T_2(0) + T_2(1) + T_1(2) + T_1(3)$. Since $T_2(1)$, $T_1(2)$, and $T_1(3)$ all are random variables, then $T_A(G)$ is also a random variable.

The ***minimum total cost C with confidence probability P under timing constraint L*** is defined as $C = \min_A C_A(G)$, where probability of $(T_A(G) \leq L) \geq P$. Probability of ($T_A(G) \leq L$) is computed by multiplying the probabilities of all nodes together while satisfying $T_A(G) \leq L$. That is, $P_A(G) = \prod_{v \in V} P_{A(v)}(v)$.

In Figure 5.2, under assignment 2, 2, 1, and 1 for nodes 0, 1, 2, and 3, $P_2(1) = Pr(T_2(0) \leq 4) = 1.0$, $P_1(2) = Pr(T_1(2) \leq 1) = 0.9$, and $P_1(3) = Pr(T_1(3) \leq 2) = 1.0$. Hence, $P_A(G) = \prod_{v \in V} P_{A(v)}(v) = 0.9$. With confidence probability P, we can guarantee that the total execution time of the graph G is less than or equal to the timing constraint L with a probability greater than or equal to P. For each timing constraint L, our algorithm will output a serial of (Probability, Cost) pairs (P, C).

$T_{R_j}(v)$ is either a discrete random variable or a continuous random variable. We define $\mathbf{F}$ to be the ***cumulative distribution function*** of the random variable $T_{R_j}(v)$ (abbreviated as ***CDF***), where $F(t) = P(T_{R_j}(v) < t)$. When $T_{R_j}(v)$ is a discrete random variable, the CDF $F(t)$ is the sum of all the probabilities associating with the execution times that are less than or equal to t. (Note: If $T_{R_j}(v)$ is a continuous random variable, then it has a

probability density function (PDF). If we assume the PDF is f, then $F(t) = \int_0^t f(s)ds$. Function F is non-decreasing, and $F(-\infty) = 0$, $F(\infty) = 1$.)

We define the ***heterogeneous assignment with probability (HAP)*** problem as follows: Given M different component types: $R_1, R_2, \cdots, R_M$, a HPDFG $G = \langle V, E \rangle$ where $V = \langle v_1, v_2, \cdots, v_N \rangle$, $T_{R_j}(v)$, $P_{R_j}(v)$, $C_{R_j}(v)$ for each node $v \in V$ executed on each component type j, and a timing constraint L, find an assignment for G that gives the *minimum total cost C with confidence probability P under timing constraint L*. In Figure 5.2, a solution to the HAP problem with timing constraint 8 can be found as follows. Assigning component types 2, 2, 1, and 1 for nodes 0, 1, 2, and 3, respectively, we obtain minimum total cost 26 with 0.9 probability under the timing constraint 8.

5.4 The Algorithm to Optimize Heterogeneous Embedded Systems

5.4.1 Definitions and Lemma

To solve the heterogeneous embedded systems optimization problem, especially configurable platform, non-fixed execution time model, we use dynamic programming method traveling the graph in a bottom up fashion. For the ease of explanation, we will index the nodes based on bottom up sequence.

Given the timing constraint L, a HPDFG G, and an assignment A, we first give several definitions as follows:

1. G^i: The sub-graph rooted at node v_i, containing all the nodes reached by node v_i. In our algorithm, each step will add one node which becomes the root of its sub-graph.

2. $C_A(G^i)$ and $T_A(G^i)$: The total cost and total execution time of G^i under the assignment A. In our algorithm, each step will achieve the minimum total cost of G^i with computed confidence probabilities under various timing constraints.

3. In our algorithm, table $D_{i,j}$ will be built. Each entry of table $D_{i,j}$ will store a linked list of (Probability, Cost) pairs sorted by probability in an ascending order. Here we define the *(Probability, Cost) pair ($P_{i,j}$, $C_{i,j}$)* as follows: $C_{i,j}$ is the minimum cost of $C_A(G^i)$ computed by all assignments A satisfying $T_A(G^i) \leq j$ with probability $\geq P_{i,j}$.

We introduce the *operator* "$\oplus$" in this chapter. For two (Probability, Cost) pairs Q_1 and Q_2, if Q_1 is $(P^1_{i,j}, C^1_{i,j})$, and Q_2 is $(P^2_{i,j}, C^2_{i,j})$, then after applying the $\oplus$ operation between Q_1 and Q_2, we get pair (P', C'), where $P' = P^1_{i,j} * P^2_{i,j}$ and $C' = C^1_{i,j} + C^2_{i,j}$. We denote this operation as "$Q_1 \oplus Q_2$".

$D_{i,j}$ is the table in which each entry has a linked list that stores pair

($P_{i,j}$, $C_{i,j}$) sorted by $P_{i,j}$ in an ascending order. Here, i represents a node number, and j represents time. For example, a linked list can be (0.1, 2)→(0.3, 3)→(0.8, 6)→(1.0, 12). Usually, there are redundant pairs in a linked list. We use Lemma 5.4.1 to cancel redundant pairs.

Lemma 5.4.1 *Given ($P^1_{i,j}$, $C^1_{i,j}$) and ($P^2_{i,j}$, $C^2_{i,j}$) in the same list:*

1. *If $P^1_{i,j} = P^2_{i,j}$, then the pair with minimum $C_{i,j}$ will be kept.*

2. *If $P^1_{i,j} < P^2_{i,j}$ and $C^1_{i,j} \geq C^2_{i,j}$, then $C^2_{i,j}$ will be kept.*

For example, if we have a list with pairs (0.1, 2) → (0.3, 3) → (0.5, 3) → (0.3, 4), we do the redundant-pair removal as follows: First, sort the list according $P_{i,j}$ in an ascending order. This list becomes (0.1, 2) → (0.3, 3) → (0.3, 4) → (0.5, 3). Second, cancel redundant pairs. Comparing (0.1, 2) and (0.3, 3), we keep both. For the two pairs (0.3, 3) and (0.3, 4), we cancel pair (0.3, 4) since the cost 4 is bigger than 3 in pair (0.3, 3). Comparing (0.3, 3) and (0.5, 3), we cancel (0.3, 3) since 0.3 < 0.5 while 3 ≥ 3. There is no information lost in redundant-pair removal.

Using Lemma 5.4.1, we can cancel many redundant pairs ($P_{i,j}$, $C_{i,j}$) whenever we find conflicting pairs in a list during a computation. After the ⊕ operation and redundant pair removal, the list of ($P_{i,j}$, $C_{i,j}$) has the following properties:

Lemma 5.4.2 *For any ($P^1_{i,j}$, $C^1_{i,j}$) and ($P^2_{i,j}$, $C^2_{i,j}$) in the same list:*

1. *$P^1_{i,j} \neq P^2_{i,j}$ and $C^1_{i,j} \neq C^2_{i,j}$.*

2. *$P^1_{i,j} < P^2_{i,j}$ if and only if $C^1_{i,j} < C^2_{i,j}$.*

In every step of our algorithm, one more node will be included for consideration. The information of this node is stored in local table $B_{i,j}$, which is similar to table $D_{i,j}$, but with cumulative probabilities only on node v_i. A local table stores only data of probabilities and cost of a node itself. Table $B_{i,j}$ is the local table only storing the information of node v_i. In more detail, $B_{i,j}$ is a local table of linked lists that store pair ($P_{i,j}$, $C_{i,j}$) sorted by $P_{i,j}$ in an ascending order; $C_{i,j}$ is the cost only for node v_i with timing constraint j, and $P_{i,j}$ is CDF (*Cumulative Distribution Function*) of time. The building procedures of $B_{i,j}$ are as follows. First, sort the execution time variations in an ascending order. Then, accumulate the probabilities of same type. Finally, let $L_{i,j}$ be the linked list in each entry of $B_{i,j}$, insert $L_{i,j}$ into $L_{i,j+1}$ while redundant pairs canceled out based on Lemma 5.4.1.

For two linked lists L_1 and L_2, the operation "$L_1 \oplus L_2$" is implemented as follows: First, implement ⊕ operation on all possible combinations of two pairs from different linked lists. Then insert the new pairs into a new linked list and remove redundant pairs using Lemma 5.4.1.

Algorithm 5.4.1 ***OESCP*** **Algorithm**

Require: α different hardware platforms, β different software components, a DAG with N nodes, and the timing constraint L

Ensure: An efficient component/platform assignment to MIN(C_p) with $Prob.(T \leq L) \geq P$

1. $M \leftarrow \alpha \times \beta$. There are total M types available.
2. $SEQ \leftarrow$ Sequence obtained by topological sorting all the nodes.
3. $t_{mp} \leftarrow$ the number of multi-parent nodes;
   $t_{mc} \leftarrow$ the number of multi-child nodes;
   If $t_{mp} < t_{mc}$, use bottom up approach;
   else, use top down approach.
4. If bottom up approach, use the following algorithm;
   If top down approach, just reverse the sequence.
5. $SEQ \leftarrow v_1 \rightarrow v_2 \rightarrow \cdots \rightarrow v_N$, in bottom up fashion;
   Add the cost and delay of each communication to the cost and delay of each node.
   $D_{1,j} \leftarrow B_{1,j}$;
   $D'_{i,j} \leftarrow$ the table that stored MIN(C_p) with $Prob.(T \leq j) \geq p$ for the sub-graph rooted on v_i except v_i;
   $v_{i_1}, v_{i_2}, \cdots, v_{i_W} \leftarrow$ all child nodes of node v_i;
   $W \leftarrow$ the number of child nodes of node v_i.

$$D'_{i,j} = \begin{cases} (0,0) & \text{if } W = 0 \\ D_{i_1,j} & \text{if } W = 1 \\ D_{i_1,j} \oplus \cdots \oplus D_{i_W,j} & \text{if } W > 1 \end{cases} \tag{5.1}$$

6. Computing $D_{i_1,j} \oplus D_{i_2,j}$:
   $G' \leftarrow$ the union of all nodes in the graphs rooted at nodes v_{i_1} and v_{i_2};
   Travel all the graphs rooted at nodes v_{i_1} and v_{i_2};
   If a node is a common node, then use a selection function to choose the type of a node.
7. For each k in $C_{i,k}$.

$$D_{i,j} = D'_{i,j-k} \oplus B_{i,k} \tag{5.2}$$

8. Then use the Lemma 5.4.1 to remove redundant pairs;
   $D_{N,j} \leftarrow$ a table of MIN(C_p) with $Prob.(T \leq j) \geq P$;
   Output $D_{N,L}$ and the configuration of hardware platforms and software components.

5.4.2 The OESCP Algorithm

Algorithm *OESCP* is shown in Algorithm 5.4.1. We have α different hardware platforms; and β different software components. $M \leftarrow \alpha \times \beta$, we get M types of components. Hence the two dimension configurable problem changed into one dimension component assignment problem. We exhaust all the possible assignments of multi-parent or multi-child nodes. For the input DAG, without loss of generality, assume using bottom up approach. If the total number of nodes with multi-parent is t, and there are maximum K variations for the execution times of all nodes, then we will give each of these t nodes a fixed assignment. We will exhaust all of the K^t possible fixed assignments. Algorithm *OESCP* gives the near-optimal solution when the given HPDFG is a DAG.

In Equation (5.1), $D_{i_1,j} \oplus D_{i_2,j}$ is computed as follows. Let G' be the union of all nodes in the graphs rooted at nodes v_{i_1} and v_{i_2}. Travel all the graphs rooted at nodes v_{i_1} and v_{i_2}. If a node q in G' appears for the first time, we add the cost of q and multiply the probability of q to $D'_{i,j}$. If q appears more than once, that is, q is a common node, then use a selection function to choose the type of q. For instance, the selection function can be defined as selecting the type that has a smaller execution time. The final $D_{N,j}$ we get is the table in which each entry has the minimum cost with a guaranteed confidence probability under the timing constraint j.

We have to solve the problem of common nodes in algorithm *OESCP*. The scenario is that one node appears in two or more graphs that are rooted at the child nodes of node v_i. In Equation (5.1), even if there are common nodes, we must not count the same node twice. That is, the cost is just added once; and the probability is multiplied once. Hence we use the information stored in the link list structure. It guarantees that the total cost equals the cost consumption of all visited nodes added together and the total probability equals the probabilities of all visited nodes multiplied together. If a common node has conflicting component type selection, we need to define a selection function to decide which component type should be chosen for the common node. For example, we can select the component type as the type that has a smaller execution time.

In this algorithm, we have considered the cost and delay of each communication by adding the cost and delay of each communication to each step of the computation. This algorithm likes to insert virtual nodes into the original HPDFG.

5.4.3 Complexity of Algorithm OESCP

In algorithm *OESCP*, it takes at most $O(N)$ to compute common nodes for each node for the worst-case, where N is the number of nodes. Thus, the complexity of the algorithm *OESCP* is $O(N^2 * L * M * K)$, where L is the given timing constraint, M is the maximum number of component types for each node, and K is the maximum number of execution time variation for each node. Usually, the execution time of each node is upper bounded by a constant, that is, L equals $O(N^c)$ (c is a constant). Hence, the complexity of *OESCP* is $O(N^2)$, which is polynomial.

5.5 Experiments

In this section, we conduct experiments with the OESCP algorithm on a set of benchmarks including Wave Digital filter (WDF), Infinite Impulse filter (IIR), Differential Pulse-Code Modulation device (DPCM), two dimensional filter (2D), Floyd-Steinberg algorithm (Floyd), and All-pole filter. We build a simulation framework to evaluate the effectiveness of our approach. α different hardware platform (H) types, H_1, $\cdots$, H_α, are used in the system, in which a hardware platform with type H_1 is the quickest with the highest energy consumption and a hardware platform with type H_α is the slowest with the lowest energy consumption. Similarly, β different software component (S) types, S_1, $\cdots$, S_β, are used in the system. Each task node has different energy consumptions under different combination of hardware platforms and software components. We have included the energy consumption and delay of each communication in the experiments. Due to the page limit, we don't list the detail table in this chapter.

We conducted experiments on four methods: Method 1: fixed platform, fixed execution time; Method 2: fixed platform, non-fixed execution time; Method 3: configurable platform, fixed execution time; Method 4: configurable platform, non-fixed execution time. We compare our solution with *hard HA ILP algorithm* [98], which is an optimal algorithm for fixed execution time model. In the list scheduling, the priority of a node is set as the longest path from this node to a leaf node [134]. The experiments are performed on a Dell PC Optiplex GX745 with a Pentium D processor and 3 GB memory running Red Hat Linux 9.0.

The experimental results for the four methods are shown in Table 5.4 to Table 5.6 when the number of software components is 3 ($\beta = 3$) and the number of hardware platforms (α) is 3, 4 and 5, respectively. Column "Bench." stands for the benchmarks we used in the experiments. Column "N." represents the number of nodes of each filter benchmark. Column "Med.1" to "Med.4" represents the four methods we used in the experiments. In Method 2 and 4, the guaranteed probability is 0.90. Column "E" represents the minimum total energy consumption obtained from three different algorithms. Column "% M1", "% M2", and "% M3" under "Med.4" represents the percentage of reduction in total energy consumption, compared to the Method 1, Method 2, and Method 3, respectively. The average reduction is shown in the last row of the table.

Four Methods Comparison with 3 FUs								
Bench.	N.	Med.1	Med.2 (0.90)	Med.3	Med.4 (0.90)			
		E (μJ)	E (μJ)	E (μJ)	E (μJ)	% M1 (%)	% M2 (%)	% M3 (%)
WDF(1)	4	852	577	485	378	55.6	34.5	22.1
WDF(2)	12	2587	1724	1432	1108	57.2	35.7	22.6
IIR	16	3596	2420	1956	1492	58.5	38.3	23.7
DPCM	16	3821	2511	2187	1752	54.1	30.2	19.9
2D(1)	34	7463	5086	4110	3106	58.4	38.9	24.4
2D(2)	4	978	638	524	416	57.5	34.8	20.6
MDFG1	8	2075	1518	1203	972	53.2	36.0	19.2
MDFG2	8	2040	1325	1098	855	58.1	35.5	22.1
Floyd	16	3810	2633	2116	1592	58.2	39.5	24.8
All-pole	29	6539	4219	3510	2691	57.7	36.2	23.3
Average Reduction (%)						56.8	36.0	22.3

TABLE 5.4: The comparison of total energy consumption with four methods on various benchmarks when the time constraint is 1000.

Four Methods Comparison with 4 FUs								
Bench.	N.	Med.1	Med.2 (0.90)	Med.3	Med.4 (0.90)			
		E (μJ)	E (μJ)	E (μJ)	E (μJ)	% M1 (%)	% M2 (%)	% M3 (%)
WDF(1)	4	1079	733	615	472	56.3	35.6	22.3
WDF(2)	12	3198	2089	1805	1355	57.6	35.1	24.9
IIR	16	4532	3035	2456	1858	59.0	38.8	24.3
DPCM	16	4607	3010	2618	2004	56.5	33.4	23.5
2D(1)	34	8435	5710	4619	3464	58.9	39.3	25.0
2D(2)	4	1182	782	678	501	57.6	35.9	26.1
MDFG1	8	2599	1857	1489	1183	54.5	36.3	20.6
MDFG2	8	2647	1785	1501	1124	57.5	37.0	25.1
Floyd	16	4934	3352	2716	2025	59.0	39.6	25.4
All-pole	29	7812	5245	4328	3259	57.8	37.2	23.9
Average Reduction (%)						57.5	36.8	24.2

TABLE 5.5: The comparison of total energy consumption with four methods on various benchmarks when the time constraint is 1000.

Four Methods Comparison with 5 FUs								
Bench.	N.	Med.1	Med.2 (0.90)	Med.3	Med.4 (0.90)			
		E (μJ)	E (μJ)	E (μJ)	E (μJ)	% M1 (%)	% M2 (%)	% M3 (%)
WDF(1)	4	1280	858	721	541	57.7	36.9	25.0
WDF(2)	12	3810	2545	2138	1592	58.2	37.4	25.5
IIR	16	5447	3648	3047	2218	59.3	39.2	27.2
DPCM	16	5583	3720	3126	2385	57.3	35.9	23.7
2D(1)	34	11320	7648	6382	4621	59.2	39.6	27.6
2D(2)	4	1428	948	825	599	58.1	36.8	27.4
MDFG1	8	3119	2192	1736	1342	57.0	38.8	22.7
MDFG2	8	3210	2210	1878	1361	57.6	38.4	27.5
Floyd	16	5911	4012	3325	2417	59.1	39.8	27.3
All-pole	29	9543	6345	5226	3945	58.7	37.8	24.5
Average Reduction (%)						58.2	38.1	25.8

TABLE 5.6: The comparison of total energy consumption with four methods on various benchmarks when the time constraint is 1000.

The results show that our algorithm OESCP can significantly improve the performance of embedded systems with multiple software/hardware components. Our algorithm for configurable platform, non-fixed execution time model improves the energy reduction over the traditional approaches. Also, we can see that with more hardware platform selections, the reduction ratio for the total energy consumption has increased. For example, with three hardware platforms, compared with Method 1, OESCP for Method 4 shows an average 56.8% reduction in total energy consumption. While using five hardware platforms, the reduction rate changed to be 58.2% for total energy consumption.

It is worthwhile to point out that we obtain this improvement ratio without sacrificing performance. For example, Table 5.7 shows the energy consumption vs. time constraints by benchmark IIR filter. OESCP for Method 4 improves the energy reduction significantly when the time constraint is large. If the time constraint is small, it still improves the energy reduction while meeting the constraint.

In conclusion, our algorithm has three main advantages: First, we design embedded systems with the consideration of configurable hardware platforms. Second, we exploit the soft real-time and non-fixed execution time to improve energy-saving. Third, we integrate these two considerations into OESCP algorithm to provide an overall energy optimization without sacrificing performance.

Four Methods Comparison on Floyd with 5 FUs							
Time	Med.1	Med.2 (0.90)	Med.3	Med.4 (0.90)			
	E (μJ)	E (μJ)	E (μJ)	E (μJ)	% M1 (%)	% M2 (%)	% M3 (%)
300	6917	5811	5378	4254	38.5	26.8	20.9
400	6623	5287	4815	3722	43.8	29.6	22.7
500	6511	4905	4450	3360	48.4	31.5	24.5
600	6409	4812	4239	3166	50.6	34.2	25.3
700	6334	4765	4077	3021	52.3	36.6	25.9
800	6163	4520	3787	2780	54.9	38.5	26.6
900	5985	4174	3481	2538	57.6	39.2	27.1
1000	5911	4012	3325	2417	59.1	39.8	27.3
1100	5789	3914	3234	2344	59.5	40.1	27.5
1200	5638	3806	3138	2272	59.7	40.3	27.6

TABLE 5.7: The comparison of total energy consumption with four methods on Floyd filter under different time constraints.

5.6 Conclusion

Since the performance of the embedded systems depends on the appropriate configuration and efficient use of different components of the whole systems, we need to consider both computation and communication of the embedded systems. In this chapter, we have considered both the energy cost and delay of each communication in a configurable hardware platform. We propose the OESCP algorithm with these two considerations to provide an overall cost optimization without sacrificing performance.

5.7 Glossary

Component-Based Software Engineering: A branch of software engineering which emphasizes the separation of concerns in respect to the wide-ranging functionality available throughout a given software system.

Computational Complexity: A measure classifying computational problems according to their inherent difficulty.

Configurable Platform Model: A model in which the hardware platform is configurable.

Fixed Platform Model: A model in which the hardware platform is fixed.

Chapter 6

Scheduling for Phase Change Memory with Scratch Pad Memory

6.1 Introduction

To close the processor-memory speed gap, the memory system of *Multiprocessor System-on-Chip* (MPSoC) is always organized as a two-level memory hierarchy: on-chip and off-chip memory. On-chip memory can be directly accessed by processors with very low latency, whereas the access latency of off-chip memory is much higher. In modern embedded processors, Scratch Pad Memory (SPM) is increasingly being employed due to its inherent advantages in terms of chip size, energy-efficiency, and timing predictability compared to cache. SPM consists of an SRAM array and decoders, which can be easily integrated into the chip. The main difference between SPM and cache is that the SPM guarantees a single-cycle access time, whereas an access to the cache is subject to cache miss which may take thousands of cycles. Data storage onto the SPM is not automatically controlled by hardware. Therefore, a scratch pad memory has 34% smaller area and 40% lower power consumption than a cache memory of the same capacity [18]. Commercial embedded processors, such as Motorola MCore [3], Texas Instruments TMS370Cx [4], Motorola 68HC12 [1], etc., take SPM as their on-chip memory.

Low energy consumption is another important issue in the embedded systems. The main memory has become one of the primary energy-consuming parts of the embedded systems. Approaches, such as [169], have been proposed to reduce the energy consumption of the mainstream DRAM main memory. Meanwhile, new techniques have been studied for the replacement of the DRAM main memory [242, 171, 113]. *Phase-Change Memory* (PCM) is a potential alterative of the DRAM main memory, due to its low energy consumption characteristic [246].

Another major advantage of the PCM over the DRAM main memory is the high scalability of the PCM memory, compared to the DRAM main memory. In a DRAM main memory, the DRAM controller must place charge the storage capacitor while mitigating sub-threshold charge leakage through the access device. To reliably sense the store charge and control the channel, both the capacitors and the transistors should be significantly large [113]. Thus, there is no manufacturable solution for scaling DRAM beyond 40nm.

On the other hand, resistive memories, such as the PCM memory, spin-torque transfer (STT) magnetoresistive RAM (MRAM), and ferroelectric RAM (FRAM), use the atomic structure changes caused by electrical current to record the content of a bit. These kinds of changes can be read by detecting the resistance. These resistive memories are easy to scale due to the fact that precise charge placement is not necessary [114]. Among the resistive memories, the PCM is the closest one to be realized and deployed [114].

To deploy the PCM memory as the main memory, the most critical issue is the endurance of the PCM memory. Writes are the primary wear mechanism in the PCM memory. As mentioned, the PCM memory places charge to change the phase of the phase-change material for a significant long time. Thus, after 10^4 to 10^9 times of writes are performed, the PCM memory cannot be programmed reliably [114]. Memory activities that reduce techniques can prolong the lifetime of the PCM memory and reduce the impact of PCM endurance issues on the reliability of an embedded system.

In many modern embedded system architecture, a small on-chip memory, which is controlled by the software, application programs or compiler support, is employed to improve both the performance and the energy efficiency of the off-chip communication, SPM. SPM consists of an SRAM array and decoders, which can be easily integrated into the chip. The main difference between SPM and cache is that the SPM guarantees a single-cycle access time, whereas an access to the cache is subject to cache miss which may take thousands of cycles. Data storage onto the SPM is not automatically controlled by hardware. Therefore, a scratch pad memory has 34% smaller area and 40% lower power consumption than a cache memory of the same capacity [18]. Commercial embedded processors, such as Motorola MCore [3], Texas Instruments TMS370Cx [4], etc., take SPM as their on-chip memory.

Since there is no hardware-controlled mechanism to transfer data between the SPM and off-chip memory, such transfers need to be explicitly managed by the software (compiler and/or application). Kandemir et al. [104] proposed a tiling-like transformation for data exchange between on-chip and off-chip memory. Xue et al. [233] introduced a data prefetching technique to hide the memory latency for accessing off-chip memory. Other previous works, such as [51, 48, 225] are also on the data transfer problem. In this paper, our target on-chip memory is organized as a *Virtually Shared SPM* (VS-SPM) architecture [103], where access to remote SPMs cost many more clock cycles than to local SPMs. We focus on data placement on the VS-SPM, and our goal is to maximize the locality of data references. So far, this issue has not been studied sufficiently.

For data-intensive applications, such as multimedia and DSP applications, the memory access operations account for approximately half of the cycle count [203], so optimizing the memory access is critical for performance improvement. For VS-SPM memory architectures, assigning variables to the appropriate SPMs will significantly reduce the overhead of remote SPM accesses. However, conventional task scheduling methods are computation-centric and ignore the effects of data layout. We strongly believe that jointly attacking both task scheduling and data partitioning is important in achieving an efficient solution for MPSoC with VS-SPM. However, each problem on its own is extremely complex, since variable partitioning decisions affect decisions on task scheduling and vice versa. The techniques proposed in literatures are mainly Integer Linear Programming (ILP) based techniques [199], [145]. The main deficiency of ILP-based techniques is the very long run time for large scale problems. In particular, it is not fit for compile-time optimization. Thus, our approach is to decompose the complex problem into two simpler subproblems that are solved in phase-ordered manner. First, we treat the VS-SPM as a large SPM with long access latency and use simple list scheduling to construct an initial schedule. Second, a heuristic variable partitioning algorithm is performed on the initial schedule to get a near-optimal variable partition. Then we use the variable partition to compact the schedule.

Our proposed variable partition algorithms are effective heuristics, in which the variable access frequencies, variable sizes and data dependencies are taken into account. The first heuristic is the *High Access Frequency First* (HAFF) variable partitioning algorithm. HAFF is a local greedy algorithm, which finds the current last finished processor and assigns the variables with higher access frequencies to the local SPM while meeting the constraints of the SPM capacities. When the latency of remote accesses is low (e.g., 5 clock cycles), HAFF performs well. The second heuristic is the *Global View Prediction* (GVP) variable partitioning algorithm. In GVP, the concept of a *Potential Critical Path* (PCP) is introduced to construct a novel weight matrix to measure the global benefit of a possible variable assignment. In each variable assignment step of GVP, we not only consider reducing the length of the current critical path, but also consider the future data requirements of potential critical paths. From experimental results, if the latency of remote accesses is long (e.g., 10 clock cycles), the schedule length of list scheduling with GVP is only 8.74% longer than the optimal schedule length on average. The optimal schedule is generated by exhaustive search. When the latency reduces to 5 clock cycles, the result is very close to the optimal schedule. The good performance of GVP is derived from estimating the global data requirements in advance.

In order to explore coarse-grained parallelism for loop portions of an application on MPSoC architecture, we propose a graph model *Data Flow Graph* (DFG) and a loop pipeline scheduling algorithm *Rotation Scheduling with Variable Partitioning* (RSVP). The DFG model can be used to explore parallelism across the loop body from different iterations. The task nodes and related variable information are extracted based on the profiling and static analysis of the application [211]. In RSVP algorithm, variable partitioning is integrated into the loop pipeline scheduling process. Rotation scheduling [41] is a mature approach to implementing loop pipelining [6]. Rotation implicitly uses retiming [44], which maintains the loop carried data dependencies of DFGs. Both HAFF and GVP can be easily integrated into RSVP when variable repartitioning is necessary. The experimental results show that the average schedule length generated by RSVP is 25.96% shorter than that of list scheduling with optimal variable partition.

The remainder of this chapter is organized as follows. Section 6.2 introduces basic concepts and related techniques. Section 6.3 formally describes the problem. Algorithms are proposed in Section 6.4. Experimental results and concluding remarks are provided in Section 6.5 and 6.6, respectively.

6.2 Models and Basic Concepts

In this section, we introduce the basic concepts which will be used in the later sections. First, we describe the target architecture model *MPSoC with Virtually Shared Scratch Pad Memory (VS-SPM)*. Second, we introduce the *Data Flow Graph* (DFG) model related to the variable partitioning problem. Third, we describe the retiming and rotation scheduling techniques.

6.2.1 Architecture Model

The organization of on-chip memory of our MPSoC architecture is depicted in Figure 6.1. Each processor core is tightly coupled with a local SPM, and all the SPMs of individual processors make up a virtually shared SPM [103]. With respect to a specific processor, the SPMs of other processors are referred to as remote SPMs. For example, in Figure 6.1, SPM_1 on processor P_1 is its local memory and $SPM_2 \cdots$, SPM_n are regarded as remote SPMs. This on-chip memory architecture takes good characteristics of both private and shared memory [146]. Since memory already occupies up to 70% of the chip area [133], and chip area is limited, the capacity of on-chip memory cannot be easily increased. Only the data to be used soon or very critical data can be loaded into SPMs. Other data is stored in the off-chip memory. The off-chip memory is composed of a large capacity PCM. In our model, off-chip memory is a shared memory that can be accessed by processor cores via an External Memory Interface (EMI) [147]. The data exchange between on-chip and off-chip memory is under the control of a Memory Controller. A Memory Controller is a hardware component controlled by dedicated instructions responsible for prefetching and writing back data from/to the off-chip PCM memory. This mechanism can help hide the very long communication latency of off-chip memory. The individual processors are connected by an interconnect bus. In this work, we assume all the processor cores are homogenous, however, our proposed techniques can be extended to heterogenous MPSoCs.

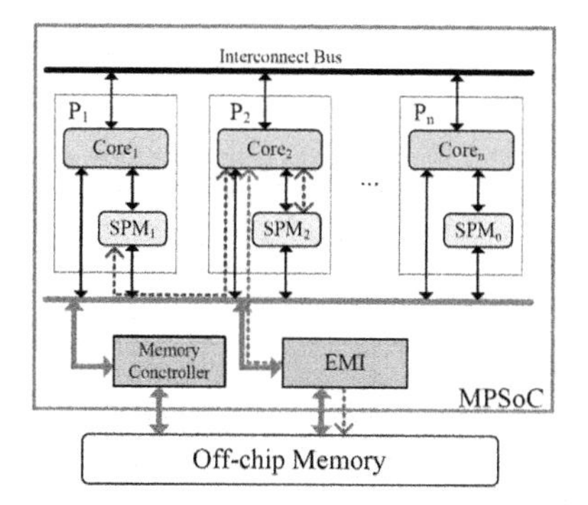

FIGURE 6.1: Architecture model MPSoC with VS-SPM.

We define three types of memory access patterns as depicted by the three dotted lines in Figure 6.1.

1. Local Access. If a processor accesses the data residing in its private SPM, the access speed is very fast. In this work, we assume local access only costs 1 clock cycle.

2. Remote Access. Due to the communication cost of the data bus, processor access to remote SPMs incurs a long access latency (e.g., 5 or 10 clock cycles).

3. Off-chip PCM Access. A processor accesses data residing in the off-chip PCM through the EMI. Because of the low performance of PCM and the high communication cost, the access latency of off-chip memory is much longer than that of SPMs.

The off-chip PCM Access latency can be hidden by data prefetching, so we only focus on variable layout on the VS-SPM in our work. For simplicity, we avoid the data coherency issue by making the assumption that a memory location can be mapped to at most one SPM. We also avoid the problem of bus access conflicts.

6.2.2 DFG Model

A task graph is a directed acyclic graph (DAG) that represents the computation blocks (tasks) of an application as nodes and communication between tasks as edges. However, a task graph cannot describe the loop carried dependencies and variable access of an application. In order to explore the pipelined loop schedule and solve the variable partitioning problem on MPSoC with VS-SPM, the *Data Flow Graph* (DFG) model is introduced in this section. In this chapter, we will use DFG as the description of a given application.

In our DFG model, each node v denotes a coarse-grain task of an application, which can be a function or a block of code. When compared to the

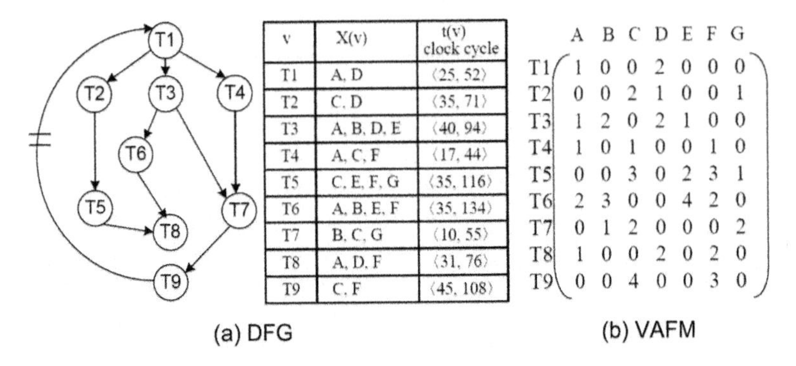

v	X(v)	t(v) clock cycle
T1	A, D	⟨25, 52⟩
T2	C, D	⟨35, 71⟩
T3	A, B, D, E	⟨40, 94⟩
T4	A, C, F	⟨17, 44⟩
T5	C, E, F, G	⟨35, 116⟩
T6	A, B, E, F	⟨35, 134⟩
T7	B, C, G	⟨10, 55⟩
T8	A, D, F	⟨31, 76⟩
T9	C, F	⟨45, 108⟩

FIGURE 6.2: Example of DFG and VAFM.

coarsely grained task nodes, communication costs are very low. Due to the high *Computation-to-Communication Ratio* (CCR), the communication overhead can be neglected in scheduling. A task consists of computation operations and memory operations (load and store). For our target MPSoC, the execution latency of computation operations are identical, and the execution latency of memory operations depend on the data layout. In other words, the execution latency of a task cannot be an exact value until data location is determined. We can only determine the execution latency bounds of a task in the DFG generation phase. The $t_{up}(v)$ and $t_{low}(v)$ are defined in Equations (6.1) and (6.2)

$$t_{up}(v) = comp_v + mem_v * R \tag{6.1}$$

$$t_{low}(v) = comp_v + mem_v \tag{6.2}$$

where $comp_v$ is the number of computation operations of task v, mem_v is the number of memory operations, R is the remote access latency. We assume each computation operation costs 1 clock cycle. Note that in this chapter we only focus on the variable placement of on-chip memory, so when all the memory operations are *Remote Access*, the corresponding latency is regarded as the upper bound task execution latency. All of these parameters can be obtained by static analysis and application profiling. The detailed DFG generation process will be introduced in Section 6.4. An example DFG is illustrated in Figure 6.2(a); we assume the remote access latency is 10 clock cycles.

Loop carried dependencies can be represented in DFGs with cycle paths. An iteration is the execution of each node in V exactly once. Inter-iteration dependencies are represented by edges with delays. The edge $e(u \rightarrow v)$ with delay count $d(e) > 0$ means that task v at the jth iteration depends on the data produced by task u at the $(j - d(e(u \rightarrow v)))^{th}$ iteration. For example, in Figure 6.2, the edge $e(T9 \rightarrow T1)$ denotes that T1 depends on the result of T9 in the last two iterations. The dependencies within the same iteration are represented by edges with zero delay. The delay count of a path is the summation of delays of all the edges along the path. Note that if all the edges with nonzero are delays removed from the DFG, the subgraph must be a DAG. A static schedule obeys the precedence relations defined by the DAG part of the DFG. For a cyclic data flow graph, the delay count of any cycle needs to be a positive integer; otherwise, the corresponding DFG is illegal.

Variable access frequencies are critical for the variable partitioning problem. According to DFG and application profiling, we construct a matrix to reflect access frequency of each variable by each task.

Definition 6.1 Variable Access Frequency Matrix (VAFM). *For a given DFG $G = \langle V, E, X, d, t\rangle$, a VAFM F is one for which $f_{i,j} \rightarrow \mathbf{Z}$, where $i \in [1, |V|]$ and $j \in [1, |X|]$. $f_{i,j}$ represents the times of the jth variable accessed by the ith task.*

The VAFM of the DFG in Figure 6.2(a) is shown as Figure 6.2(b). Each row reflects the variable accesses of one task. From the VAFM, we can get a global view of relations between tasks and variables.

6.2.3 Retiming and Rotation Scheduling

The techniques described in this section will be used in our loop pipeline scheduling algorithm RSVP. Retiming [45] has been effectively used to obtain the minimum cycle period for a DFG by rearranging the delays. It is defined as a function $r(v)$ from V to an integer. The value $r(v)$ is the number of delays drawn from each of the incoming edges of node v and pushed to each of the outgoing edges of v. Given a DFG $G = \langle V, E, X, d, t\rangle$, a legal retiming function r with cycle period c must satisfy the following conditions:

1. For every edge $e(u \rightarrow v) \in E$, $d_r(e) = d(e) + r(u) - r(v) \geq 0$. So $r(v) - r(u) \leq d(e)$ for each edge $e(u \rightarrow v) \in E$

2. For each path $u \rightsquigarrow v$, $d_r(p) = d(p) + r(u) - r(v) \geq 1$ if $t(p) > c$. So $r(v) - r(u) \leq d(p) - 1$ for each path $u \rightsquigarrow v$ if $t(p) > c$

where: $d_r(e)$ is the delay of edge e after retiming; $d(p)$ is the number of delays of path p; $t(p)$ is the computation time of path p. In Figure 6.3(a), $r(T3) = 1$ is a legal retiming, since the delay of incoming edges of $T3$ are no less than 1. If retiming function is $r(T3) = 2$, $d_r(e(T2 \leftarrow T3)) = 1 - 2 = -1 < 0$, that means $T3$ in this iteration depends on the result of $T2$ of a future iteration, so $r(T3) = 2$ is an illegal retiming. After retiming, the length of the critical path would be reduced and a more compact iteration period will be obtained. For instance, the DFG in Figure 6.3 (a), when $r(T1) = 1$ and $r(Ti) = 0, \forall i = [2 \cdots 9]$, the retimed graph G_r is shown as Figure 6.3(b). The critical path of DAG of G_r is $T3 \rightarrow T6 \rightarrow T8$, which is shorter than the original critical path $T1 \rightarrow T3 \rightarrow T6 \rightarrow T8$.

Rotation scheduling presented in [42] is a loop scheduling technique used to optimize loop scheduling with resource constraints. It transforms a schedule to a more compact schedule iteratively. In most cases, the node level minimum

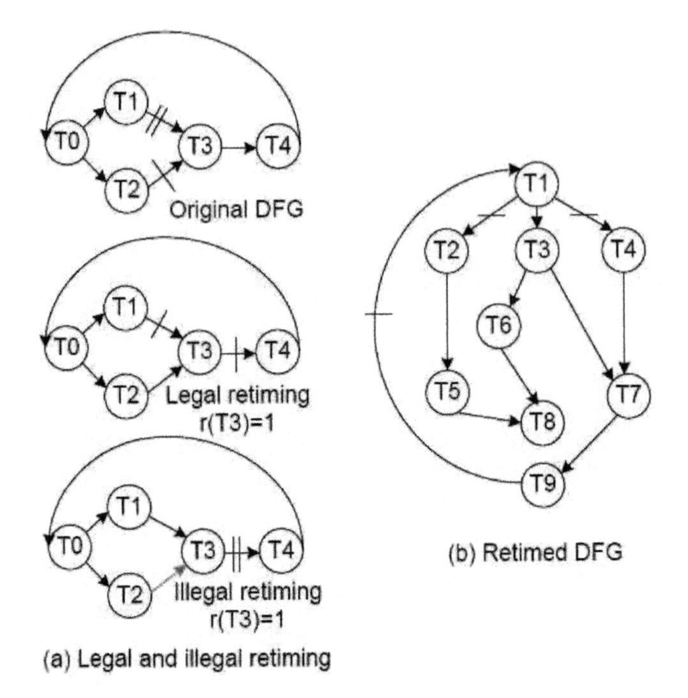

FIGURE 6.3: Retiming technology.

schedule length can be obtained in polynomial time by rotation scheduling. In each step of the rotation, nodes in the first row of the schedule are rotated down. By doing so, the nodes in the first row are rescheduled to the earliest possible available locations. From retiming point of view, each node gets retimed once by removing one delay from each of incoming edges of the node and adding one delay to each of its outgoing edges in the DFG. The details of rotation scheduling can be found in [42].

6.3 Problem Statement

Task scheduling and variable partitioning are interdependent problems. The following motivational example demonstrates the interactions of these two problems. For simplicity, we assume that there are only two parallel executed tasks. The VAFM and the number of computation operations of each task are shown in Figure 6.4(a). There are two different variable partition schemas as shown in Figure 6.4(b) and (c). The dotted lines denote remote accesses (latency is 10 clock cycles) and the solid lines denote local accesses. Comparing these two variable partitions, the execution latencies of the two tasks decrease by 49.5% and 38.1%, respectively, by the schema2. The schedule length of schema2 is only 61.9% compared to schema1.

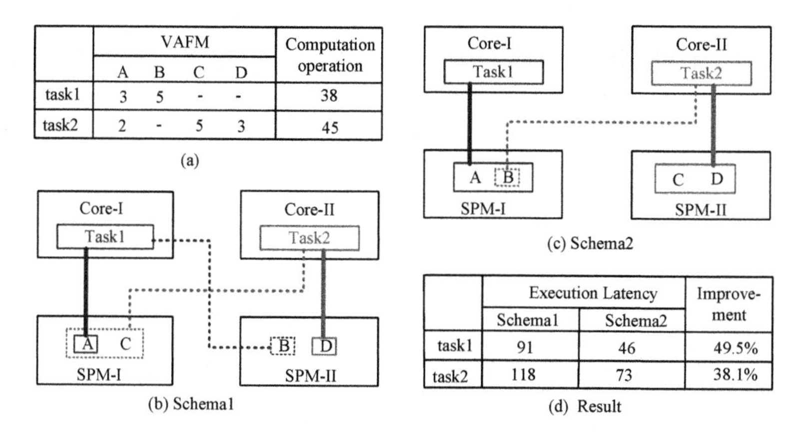

FIGURE 6.4: Interaction between variable partitioning and task scheduling.

Variable Partitioning Problem: For a given DFG $G = \langle V, E, X, d, t \rangle$ and target MPSoC M, variable assignment is a function η: $X \rightarrow M$, that maps a variable x_i onto a processor P_j. Then, the Variable Partition (VP) is defined as Equation (6.3)

$$VP = \{\langle x_i, P_j \rangle : x_i \in X, P_j \in M, \eta(x_i) = P_j\} \tag{6.3}$$

A legal partition should satisfy the constraint:

$$\sum_{\forall x_i \in \{x_i : \eta(x_i) = P_j\}} Size(x_i) \leq Msize(SPM_j) \tag{6.4}$$

where $Size(x_i)$ is the size of variable x_i. For example, in Figure 6.4(b), we assume that the capacity constraint (6.4) is satisfied, $VP = \{\langle A, P_1 \rangle, \langle B, P_2 \rangle, \langle C, P_1 \rangle, \langle D, P_2 \rangle\}$ is a legal partition of variable set $\{A, B, C, D\}$. To avoid the complexity of data consistency, one variable can only be assigned to one SPM.

Task Scheduling Problem: For a given application specified as DFG $G = \langle V, E, X, d, t \rangle$ and a target MPSoC M, task scheduling maps the tasks to processors and suggests a start time for each task while maintaining the dependencies defined by E. We define a function as $p(v_i) = P_j$, where $v_i \in V$ *and* $P_j \in M$, which represents that the ith task is mapped onto the jth processor. If a task v_i is scheduled on processor P_j, $ST(v_i, P_j)$ and $FT(v_j, P_j)$ denote the start time and finish time of v_i on processor P_j, respectively. It should be noted that $FT(v_i, P_j) = ST(v_i, P_j) + Latency(v_i, P_j)$. Since execution latency of a task depends on the variable partition, we define task latency in Equation (6.5).

$$Latency(v_i, P_j) = comp_{v_i} + l\_mem_{v_i} + r\_mem_{v_i} * R \qquad (6.5)$$

where $comp_{v_i}$ is the number of computation operations in task v_i, $l\_mem_{v_i}$ is the number of local access memory operation, $r\_mem_{v_i}$ is the number of remote access memory operation, R is the remote access latency. After all tasks have been scheduled, the schedule length SL is defined as Equation (6.6).

$$SL = \max_{\forall v_i \in V, \forall P_j \in M} \{FT(v_i, P_j)\} \qquad (6.6)$$

Goal: The goal of our algorithm is to find an appropriate variable partition and a corresponding task schedule to minimize SL. Scheduling problems with time and resource constraints are known to be NP-complete. We are going to solve this kind of scheduling problem with variable partitioning. Therefore, our problem is also NP-complete. The proposed heuristic algorithms can approximate the optimal schedule in polynomial time. They are more efficient than ILP-based solutions for large-scale problems (the number of processors, variables and tasks are large).

6.4 Algorithms

Solving variable partitioning and task scheduling problems simultaneously is extremely hard. We decouple these problems and solve them in a phase-ordered manner. In the first phase, for a given DFG, we generate an initial schedule S_{init} based on upper bound execution latencies of tasks. Any kind of heuristic algorithm can be used in the initial scheduling. For simplicity, we use list scheduling taking the number of descendants as the weight of a node. For example, considering the DFG in Figure 6.5(a), the S_{init} is generated by list scheduling shown in Figure 6.5(b). In the second phase, under the guide of the S_{init} and $VFAM$, the proposed variable partitioning algorithm will assign variables to appropriate SPMs to compact the S_{init}.

6.4.1 High Access Frequency First (HAFF) Variable Partition

The HAFF algorithm is shown in Algorithm 6.4.1. Based on the S_{init} and $VAFM$, we define a matrix (Core-Variable Matrix) to represent the access frequencies of each variable accessed by different processor cores.

Definition 6.2 Core-Variable Matrix (CVM). *For a given initial schedule S_{init}, $VAFM$ and target MPSoC M, CVM is an $|M| \times |X|$ matrix as* *follows:*

$$CVM[i,j] = \sum_{\forall v_k \in \{v_k : p(v_k) = P_i\}} VAFM[k,j]$$

where v_k is a task node, $|M|$ is the number of processor cores and $|X|$ is the number of variables. $CVM[i,j]$ represents the times of the jth variable accessed by the tasks on the ith processor core.

Algorithm 6.4.1 High Frequency First Variable Partition

Require: (1) Initial schedule: S_{init}, (2) Target machine: M, (3) DFG: G, (4) VAFM

Ensure: (1) Variable partition: VP, (2) Compact schedule: S_{comp}

1: $T \leftarrow M$;
2: $CVM \leftarrow CVM_$Generate$(VAFM, S_{init})$;
3: **while** (!Partition_finished( )) **do**
4:     $P_\alpha \leftarrow$ find a target processor from T by descending finish time order;
5:     **if** (P_α) **then**
6:         $var_\beta \leftarrow$ find a target variable by descending frequencies order, and the var_β should satisfy constraints 1) $CVM[P_\alpha, var_\beta] > 0$ and 2) FreeSpace$(SPM_\alpha) >$ Size(var_β);
7:         **if** (var_β) **then**
8:             $VP \leftarrow VP \bigcup \{\langle var, P \rangle\}$;
9:             FreeSpace$(SPM_\alpha) \leftarrow$ FreeSpace(SPM_α) −Size(var_β);
10:            $S_{comp} \leftarrow$ Schedule_Compact(S_{init}, G, VP);
11:        **else**
12:            $T \leftarrow T - P_\alpha$;
13:        **end if**
14:    **else**
15:        Assign the left unassigned variables to make full use of SPMs
16:    **end if**
17: **end while**
18: **Return** VP and S_{comp}

Step 1. In lines 1–2, we initialize T and construct the CVM. Note that by assigning variables to a processor $P_\alpha \in T$ we can compact the schedule.

Step 2. We find the target processor P_α which is to be assigned a variable (line 4). In the variable assignment process, the processor in T with the longest finish time is to be the target processor. After a variable is assigned, the finish times of processors may change. So, we will choose the target processor in each round of variable assignment. For example, in Figure 6.5(b), P_1 is the target processor in first round of variable assignment. After variable "A" has been assigned to SPM_1, P_2 becomes the target processor. Since the SPMs capacities constraint, after several rounds of variable assignment, we cannot find the P_α from T. We will assign the left variables only considering the SPMs capacities (line 15).

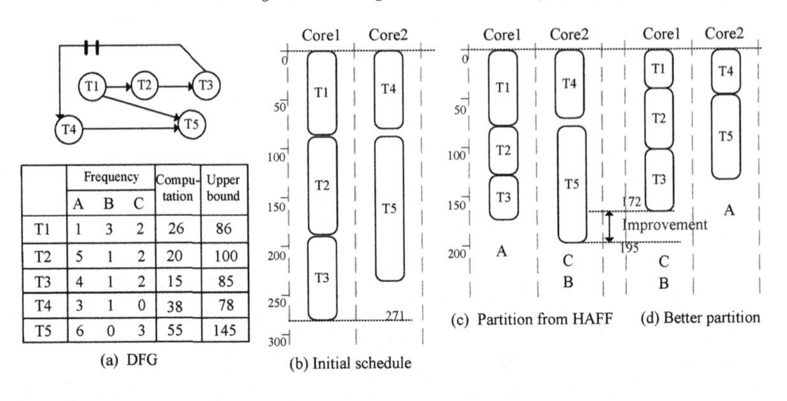

FIGURE 6.5: HAFF vs. GVP.

Step 3. After the target processor P_α is determined, we will choose a variable which has higher access frequencies by P_α and meet the capacity constraint as the target variable var_β (line 6). For our example as Figure 6.5, in the first round variable assignment P_1 is the target processor, variable "A" is the target variable, because it has the highest access frequencies by P_1. If we cannot find the target variable var_β for P_α, then the P_α will not be the target processor in future rounds of variable assignment (line 12).

Step 4. If a variable assignment $\langle var_\beta, P_\alpha \rangle$ is found in step 3, we will add it to the variable partition VP and mark the var_β as an assigned variable. We also update the free space of SPM_α. Then, we call "Schedule_Compact" to generate a more compact schedule. The "Schedule_Compact" process compacts the latency of related tasks based on the current VP to generate a shorter schedule. Note that "Schedule_Compact" keeps the relative start order in S_{init} while maintaining the dependencies defined in DFG. The pseudo-code of this step is shown in lines 8–10.

The "Partition_finished( )" is determination condition, when all the variables in VAFM are assigned or the left space of SPMs are smaller than any unassigned variables. In Algorithm 6.4.1, we sort the processors by finish time and the variables by access frequencies to determine the P_α and var_β. The time complexity of the two sorts are $\Theta(|M|log|M|)$ and $\Theta(|X|log|X|)$, respectively, where M is number of processors and $|X|$ is number of variables. Thus, the overall time complexity of HAFF algorithm is $\Theta(|X||M|log|M| + |X|^2log|X|)$.

6.4.2 Global View Prediction (GVP) Variable Partition

In the HAFF algorithm, the processors with longer finish times have higher priorities to be assigned a variable with a higher access frequency. According to our experimental results, HAFF can often generate a near-optimal schedule. However, sometimes to compact the overall schedule length the var_β (in Algorithm 6.4.1) should not be assigned to the P_α. The example in Figure 6.5

illustrates this limitation of HAFF. For the DFG and S_{init} in Figure 6.5(a) and (b), we assume that remote access latency is 10 clock cycles. The variable partition generated by HAFF is: $VP_1 = \{\langle A, P1\rangle, \langle B, P2\rangle, \langle C, P2\rangle\}$. Based on VP_1, the schedule length has been compacted to 195 control steps as shown in Figure 6.5(c). It is 71.9% of the initial schedule, which shows that VS-SPM architecture is much better than shared SPM architecture. However, we can obtain a better schedule shown in Figure 6.5(d) from another variable partition $VP' = \{\langle A, P2\rangle, \langle B, P1\rangle, \langle C, P1\rangle\}$. The schedule length based on this variable partition is only 172 control steps, 11.9% shorter than the schedule based on VP.

From the above example, we know that the main limitation of HAFF is caused by its local greedy strategy. Future variable requirements are not taken into account when assigning each variable. In this section, we propose a novel heuristic algorithm called Global View Prediction (GVP) variable partition shown in Algorithm 6.4.2. The main difference between GVP and HAFF is that GVP has the capability of predicting the variable requirements of potential target processors for global optimization.

Step 1. We construct a new directed graph $DAG_{vp} = (V', E')$ based on the given DFG and its initial schedule S_{init}, where V' consists of node set of DFG V and dummy head/tail nodes, E' represents the execution order of tasks in S_{init}. For example, the Figure 6.6(a) is the DAG_{vp} generated from DFG and S_{init} in Figure 6.5(a) and (b). $e(T1 \rightarrow T5)$ reflects the precedence of tasks on different cores, $T5$ cannot start until $T1$ finished. The dummy nodes facilitate finding the critical path and the paths through a specific node.

Step 2. We find the Current Critical Path (CCP), Potential Critical Path (PCP) of DAG_{vp} and variables access matrix (CVM_{ccp} and CVM_{pcp}) related to these paths (line 4–6). From Equation (6.5) we know that the length of tasks are uncertain until the variable partition is determined. So, we find the CCP and PCP while keeping in mind that some variables have already been assigned, which is called the current variable partition.

Definition 6.3 (CCP and PCP). *For a given DAG_{vp} and current variable partition VP, Current Critical Path CCP is the longest path from head to tail. A Potential Critical Path PCP from head to tail satisfies the condition: $Length(PCP) > Length(CP_{lb})$, where $Length(PCP)$ is the length of PCP, $Length(CP_{lb})$ is the length of critical path when all task execution latencies are t_{low}. PS denotes the set of all PCPs.*

For example, the CCP of DAG_{vp} in Figure 6.6 is $(head \rightarrow T1 \rightarrow T2 \rightarrow T3 \rightarrow tail)$. Usually, the number of potential critical path is larger than one. We will consider all the paths in PS. One variable is assigned to an SPM in each round of variable assignment. The original critical path may be compacted after some rounds, then one or more PCPs become the CPP. For example, in Figure 6.6(b) the "PathII" $(head \rightarrow T1 \rightarrow T5 \rightarrow tail)$ and "PathIII" $(head \rightarrow T1 \rightarrow T4 \rightarrow tail)$ are the potential critical paths. If variable "A" is assigned to SPM_1 then "PathII" will become the CCP.

Algorithm 6.4.2 Global View Prediction Variable Partition

Require: (1) Initial schedule: S_{init}, (2) Target machine: M, (3) DFG: G (4)VAFM

Ensure: (1) variable partition: VP, (2) compact schedule: S_{comp}

1: $DAG_{vp} \leftarrow$ Generate_$DAG_{vp}(S_{init}, G)$;
2: Initialize λ_1 and λ_2;
3: **while** (!Partition_finished( )) **do**
4: $\quad CCP \leftarrow$ Find_CCP(DAG_{vp});
5: $\quad PS \leftarrow$ Find_PCP(DAG_{vp});
6: $\quad (CVM_{ccp}, CVM_{pcp}) \leftarrow (CCP, PS, VAFM)$;
7: $\quad CP_gain[i,j] \leftarrow SL(S_{init})-$Schedule_Compact$(S_{init}, VP \bigcup \{\langle var_i, P_j\rangle\})$;
8: $\quad Global\_gian[i,j] \leftarrow \sum_{p(task_k)=P_j} VAFM(var_i, P_j) * PCP\_through(task_k)$;
9: $\quad Weight[i,j] \leftarrow \lambda_1 * CP\_gain[i,j] + \lambda_2 * Global\_gain[i,j]$;
10: $\quad$ **if** (find a legal assignment based on $Weight$) **then**
11: $\quad\quad VP \leftarrow VP \bigcup \{\langle var, P\rangle\}$;
12: $\quad$ **else**
13: $\quad\quad IBN\_gain \leftarrow$ generate an IBN_gain;
14: $\quad\quad$ **if** (find a legal assignment with higher value in IBN_gain) **then**
15: $\quad\quad\quad VP \leftarrow VP \bigcup \{\langle var, P\rangle\}$;
16: $\quad\quad$ **else**
17: $\quad\quad\quad$ find a legal variable assignment and try to make full use of SPMs
18: $\quad\quad\quad VP \leftarrow VP \bigcup \{\langle var, P\rangle\}$;
19: $\quad\quad$ **end if**
20: $\quad$ **end if**
21: $\quad$ FreeSpace(SPM_P) $\leftarrow$ FreeSpace(SPM_P) - Size(var);
22: $\quad S_{comp} \leftarrow$ Schedule_Compact(S, G, VP);
23: $\quad \lambda_1 \leftarrow \lambda_1 + \delta$;
24: $\quad \lambda_2 \leftarrow 1 - \lambda_1$;
25: **end while**
26: Return VP and S_{comp}

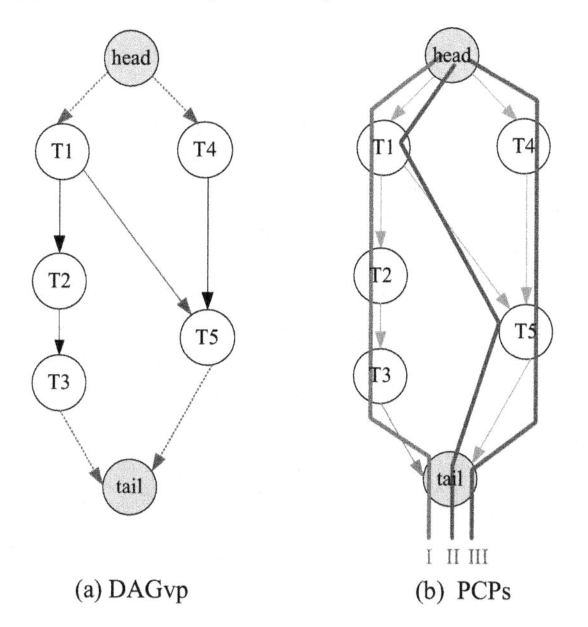

(a) DAGvp (b) PCPs

FIGURE 6.6: DAG_{vp} and potential critical path.

Definition 6.4 ($\mathbf{CVM}_{ccp}$ and $\mathbf{CVM}_{pcp}$). *For a given DAG_{vp}, CCP, and PS, CVM_{ccp} and CVM_{pcp} are two $|M| \times |X|$ matrixes shown as Equations (6.7) and (6.8)*

$$CVM_{ccp}[i,j] = \begin{cases} 1, & \textbf{if } \forall task_k \in CCP,\ p(task_k) = P_i \\ & \quad and\ VAFM[task_k, x_j] > 0 \\ 0, & \textbf{otherwise} \end{cases} \tag{6.7}$$

$$CVM_{pcp}[i,j] = \begin{cases} 1, & \textbf{if } \forall task_k \in PS,\ p(task_k) = P_i \\ & \quad and\ VAFM[task_k, x_j] > 0 \\ 0, & \textbf{otherwise} \end{cases} \tag{6.8}$$

The $CVM_{ccp}[i,j] = 1$ represents that the variable assignment $\langle var_j, P_i \rangle$ will improve the schedule length. Similarly, $CVM_{pcp}[i,j] = 1$ represents that it is possible to improve the schedule length by the variable assignment $\langle var_j, P_i \rangle$. For our example in Figure 6.5, in the first round of variable assignment, the CVM_{pcp} is a 2×3 one matrix and

$$CVM_{ccp} = \begin{pmatrix} 1 & 1 & 1 \\ 0 & 0 & 0 \end{pmatrix}$$

Step 3. A matrix $Weight$ is constructed to measure the schedule length improvements from a variable assignment $\langle var_i, P_j \rangle$. The pseudo-code of this step is shown in lines 7–9. $Weight$ is computed from two $|M| \times |X|$ matrices CP_gain and $Global\_gain$.

Definition 6.5 (CP_gain). *Given a DAG_{vp} and a VAFM, a CP_gain matrix is one for which*

$$CP\_gain[i,j] = \begin{cases} SL_{init} - SL_{i,j}, & \textbf{if } CVM_{ccp}[i,j] = 1 \\ 0, & \textbf{otherwise} \end{cases} \tag{6.9}$$

where SL_{init} is the length of initial schedule and $SL_{i,j}$ is the length of initial schedule compacted by $VP \bigcup \{\langle var, P\rangle\}$.

CP_gain matrix reflects the improvements of the schedule length after possible variable assignments are applied. For example, in Figure 6.5(a), the CP_gain of first round variable assignment is

$$CP\_gain = \begin{pmatrix} 49 & 48 & 48 \\ 0 & 0 & 0 \end{pmatrix}$$

$CP\_gain[1][1]$=49 means when variable "A" is assigned to SPM_1, the schedule length will be compacted by 49 control steps. If we only use CP_gain as the weight to determine a variable assignment, it is still a local greedy strategy. So, we will consider all the potential critical paths of DAG_{vp} and get a $Global\_gain$ weight matrix.

Definition 6.6 Global_gain Matrix. *Given a DAG_{vp} and related VAFM, a Global_gain matrix is one for which*

$$Global\_gain[i,j] = \begin{cases} \sum_{p(task_k)=P_i} VAFM(task_k, x_j) * PCP\_through(task_k), & \textit{if } CVM_{pcp}[i,j]=1 \\ 0, & \textit{otherwise} \end{cases}$$

where $PCP\_through(task_k)$ represents the number of potential critical paths that pass through the node $task_k$.

For example, in Figure 6.6(b), $PCP\_through(T1) = PCP\_through(T5)$=2, and $PCP\_through(i) = 1, \forall i \in \{T2, T3, T4\}$. The $Global\_gain$ of the first variable assignment round is:

$$Global\_gain = \begin{pmatrix} 11 & 8 & 8 \\ 15 & 1 & 6 \end{pmatrix}$$

$Global\_gain$ is global view measurement of this round variable assignment, which reflects the potential improvements from this assignment. To reflect the tradeoff between CP_gain and $Global\_gain$, we use a weighted sum formula to compute the $Weight$ matrix. The average weight of a variable assignment $Weight[i,j]$ is defined as

$$Weight[i,j] = \lambda_1 * CP\_gain[i,j] + \lambda_2 * Global\_gain[i,j] \tag{6.10}$$

where $\lambda 1, \lambda 2$ are two coefficients that represent the weight of CP_gain and $Global\_gain$. So, the $Weight$ provides a complete measurement of benefit obtained from a candidate variable assignment. In the beginning of variable partition, we should pay more attention to global benefit of a variable assignment, such as $\lambda_1 = 0.1, \lambda_2 = 0.9$. After some variables have been assigned, the number of possible variable assignments decreases rapidly, so the *Global_gain* will be less important. The two coefficients are defined as follow:

$$\lambda_1[i] = \lambda_1[i-1] + \delta,$$
$$\lambda_2[i] = 1 - \lambda_1[i],$$

where $\delta = \frac{1}{2*|X|}$. Note, in order to ensure the elements of CP_gain and $Global\_gain$ are in the same range, each member of the CP_gain and $Global\_gain$ matrices is normalized into [0,1] with the formula $CP\_gain[i,j] = \frac{CP\_gain[i,j]}{max_{i,j}\{CP\_gain[i,j]\}}$ and $Global\_gain[i,j] = \frac{Global\_gain[i,j]}{max_{i,j}\{Global\_gain[i,j]\}}$, respectively. In each variable assignment round, the variable assignment that has the largest weight and satisfies the memory space constraint is added into the current variable partition VP. For example, in the first round of variable assignment of Figure 6.5(a), the $Weight$ matrix is as follows:

$$\begin{aligned} Weight &= 0.17 * \begin{pmatrix} 1 & 0.97 & 0.97 \\ 0 & 0 & 0 \end{pmatrix} + 0.83 * \begin{pmatrix} 0.73 & 0.53 & 0.53 \\ 1 & 0.07 & 0.4 \end{pmatrix} \\ &= \begin{pmatrix} 0.77 & 0.44 & 0.44 \\ 0.83 & 0.06 & 0.33 \end{pmatrix} \end{aligned}$$

The variable assignment $\langle \text{"}A\text{"}, P_2 \rangle$ is the result of first variable assignment round.

Step 4. After several rounds, due to the SPM capacity constraint, legal assignments cannot be found based on the $Weight$ matrix. We will consider other variable assignment options. The pseudo-code of this step is shown in lines 13–18.

Definition 6.7 CPN, IBN and OBN. *A* Critical Path Node *(CPN) is a node on the critical path. An* In-Branch Node *(IBN) is a node, from which there is a path reaching a CPN. An* Out-Branch Node *(OBN) is a node, which is neither a CPN or an IBN.*

Compacting the latency of CPNs and IBNs is beneficial to compacting the schedule length. If there are no legal assignments related to CPNs, we will consider the assignments related to IBNs. Similar to CP_gain, we construct a matrix IBN_gain which represents the schedule length improvement from the variable assignment related to an IBN. Due to space limitations, we omit the definition IBN_gain. We try to find the legal variable assignment with the largest value in IBN_gain to be the result assignment of this round. When the available SPM space of processors related to CPNs and IBNs cannot accommodate any more variables, we will consider all other legal assignment options. In this case, we try to make full use of left over space of SPMs.

Step 5. When the variable assignment $\langle var, P \rangle$ is determined, the variable partition VP and free space of SPM_P are updated. The weight coefficients λ_1, λ_2, are updated as well. When all the variables have been assigned or the left space of SPMs are not enough for any variables, the variable partition VP and compact schedule S_{comp} are returned.

The two most time intensive portions of the GVP algorithm are path searching and weight matrix construction. The time complexity of CCP and PCP path search is $\Theta(|V|^2)$. The time complexity of weight matrix construction is $\Theta(|V||X||M|)$, where $|V|$ is number of task nodes, $|X|$ is number of variables, $|M|$ is number of processors. Thus, the overall time complexity of GVP algorithm is $\Theta(|X||V|^2 + |X|^2|V||M|)$.

6.4.3 Loop Pipeline Scheduling

In the previous section, list scheduling with variable partition for the acyclic part of DFG has been introduced. Loops are usually the most time-critical portions of data-intensive applications. In order to improve overall throughput, loop pipeline scheduling — an optimization technique for cyclic DFG [6] — needs to be exploited. This section discusses the proposed loop pipeline scheduling algorithm *Rotation Scheduling with Variable Partition* (RSVP) as shown in Algorithm 6.4.3. RSVP algorithm can produce a compact schedule by repeatedly applying retiming (introduced in Section 6.2.3) and remapping the rotated down nodes to new positions on the schedule table. The novel feature of RSVP algorithm is that it takes variable partitioning into account. During loop pipeline scheduling, variable partitioning is executed only a few times when it is necessary.

Step 1. The initial schedule and variable partition are generated in this step. It is shown in lines 1–3 of Algorithm 6.4.3. S_{opt} is the reference schedule.

Lines 5–19 of Algorithm 6.4.3 is one round of rotation. We try to take some nodes from the first row of a schedule table and remap them to new positions while keeping the remaining nodes in their original schedule spots. The variable partition must be taken into account in the node remapping process.

Step 2. We find out the rotatable node set *RotSet* and rotate these nodes down (lines 5–6). A rotatable node is a node in the first row of current schedule table S and the delays of its coming edges are larger or equal to one in the DFG. Then, we rotate down the nodes in *RotSet* by the retiming technique explained in Section 6.2.3. This operation is equivalent to shifting the iteration boundary down so that the nodes from the next iteration can be explored. After that, we get a new DFG G_r. For example, Figure 6.7(a) and (b) illustrate the first round of RSVP for the DFG in Figure 6.2. The rotatable nodes set is $RotSet = \{T1\}$. After retiming $r(T1) = 1$, the DFG in Figure 6.2(a) is transformed into G_r as shown in Figure 6.3(b).

Step 3. Remap the nodes in *RotSet* to appropriate positions of the new schedule table. There are two factors that need to be taken into account in remapping. First is the dependency constraints in the DAG part of G_r. Second is current variable layout. The pseudo-code of the remapping is shown in Algorithm 6.4.4. Since the latency of tasks are different, when the rotatable

Algorithm 6.4.3 Rotation Scheduling with Variable Partition

Require: (1) DFG: G, (2) VAFM: F, (3) MPSoC: M
Ensure: (1) near-optimal variable partition: VP_{opt}, (2) near-optimal schedule: S_{opt}

1: $S \leftarrow$ list_scheduling(G, M);
2: $(S, VP) \leftarrow$ GVP(S, F); { HAFF can be used as well}
3: $S_{opt} \leftarrow S$;
4: **for** $i = 1$ to $\mathcal{Z}$ **do**
5: $RotSet \leftarrow$ find rotatable nodes;
6: $G_r \leftarrow$ retiming(G, $RotSet$);
7: **if** $G_r = G$ **then**
8: **break**; {If G_r is identical G, it is not necessary to continue rotation}
9: **end if**
10: $S_{rot} \leftarrow$ remapping($RotSet, VP, S, F$);
11: **if** $SL(S_{rot}) > SL(S)$ **then**
12: $S' \leftarrow$ list_scheduling(G_r, M); {The upper bound of execution latencies are used}
13: $(S_{rot}, VP) \leftarrow$ GVP(S', M, G, F); {HAFF or other algorithms can be used for repartitioning}
14: **end if**
15: $S \leftarrow S_{rot}$;
16: **if** $SL(S) < SL(S_{opt})$ **then**
17: $S_{opt} \leftarrow S$;
18: $VP_{opt} \leftarrow VP$;
19: **end if**
20: **end for**
21: Return S_{opt} and VP_{opt}

nodes are rotated down, there may be some different idle time slots in the beginning of some processors. Before the nodes are remapped, we reschedule the nodes in $\{V - RotSet\}$ to fill up those idle time slots. Note that the processor allocation and relative start order of these nodes are not changed in this rescheduling.

The basic principle of remapping is to make the finish times of a rotatable task as soon as possible. To obtain shorter schedule lengths, the tasks with longer t_{up} have higher priority to be remapped first (line 2 of Algorithm 6.4.4). The processor on which h can finish first is the target processor for remapping h (lines 6–11 of Algorithm 6.4.4). After all the rotatable nodes have been remapped, a new schedule is returned. For example, in Figure 6.7(b) is a new schedule where task 1 has been mapped to Core 4. Even though the latency of task 1 increases, the schedule length is compacted to 251 control steps.

Step 4. The pseudo-code of this step is shown in lines 11–14 of Algorithm 6.4.3. We compare the schedule length of S_{rot} $SL(S_{rot})$ with the length of last round schedule $SL(S)$ (line 11). If the $SL(S_{rot})$ is longer than $SL(S)$, we will generate a new schedule S' for G_r by list scheduling and use GVP (Algorithm 6.4.2) to repartition all the variables. Note that the increase of $SL(S_{rot})$ is not only caused by the variable partition, but also by the length of critical

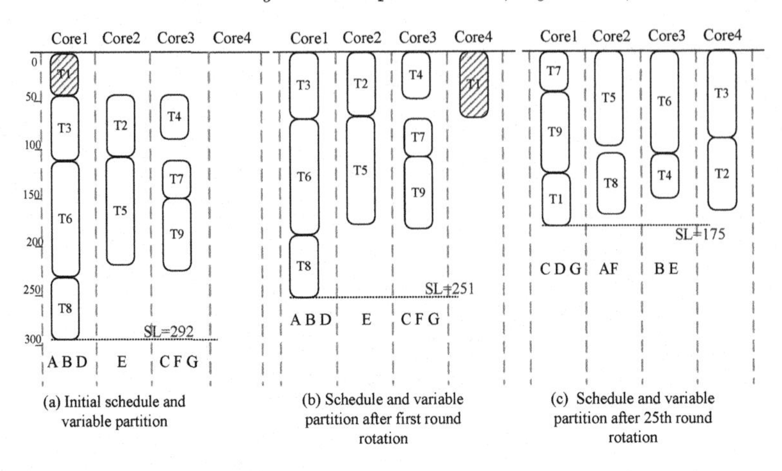

(a) Initial schedule and variable partition
(b) Schedule and variable partition after first round rotation
(c) Schedule and variable partition after 25th round rotation

FIGURE 6.7: Rotation scheduling with variable partition.

path which is varied by retiming. Even though variable repartitioning and rescheduling are executed, the length of S_{rot} may be still longer than reference length. RSVP algorithm continues iterating on this new state (S_{rot}) instead of returning to the previous schedule state in hopes of reducing the schedule length in future iterations.

Step 5. At the end of each iteration, the best schedule and corresponding variable partition among all previous rotation rounds are saved.

The first termination condition is an argument $\mathcal{Z}$, which restricts the number of rotation iterations. When the algorithm reaches this value, the shortest schedule and corresponding variable partition are returned. $\mathcal{Z}$ can be any integer value, but according to experimental results we define $\mathcal{Z}$ as Equation (6.11) to get near-optimal schedule.

$$\mathcal{Z} = \frac{SL(S_{init})}{min\{t_{up}(task_i)\}} * |M|, \tag{6.11}$$

where SL(S_{init}) is schedule length of initial schedule, $|M|$ is number of processors in the MPSoc. Using the example in Figure 6.2, after the 25th round of rotation, we obtain a very compact pipelined schedule shown in Figure 6.7 (c), which is only 59.9% the length of the un-pipelined schedule. Note that after several rounds of rotation, G_r may be the same as the original input G. If this happens, it is not necessary to do any more rotation rounds. We check for this termination condition in line 7.

Algorithm 6.4.4 Remapping

Require: (1) a rotatable node set: $RotSet$, (2) variable partition: VP, (3) schedule: S, (4)VAFM: F, (5) retimed DFG: G_r
Ensure: a new schedule: S_{rot}
1: reschedule the note in $\{V - Rotset\}$ to fill up the idle time slots
2: $list_{rp} \leftarrow$ generate an ordered list by descending order of t_{up} ;
3: $h \leftarrow$ head of $list_{rp}$;
4: **while** ($h \neq$ NULL) **do**
5: $t_{rp} \leftarrow \infty$;
6: **for** $i = 1$ to $|M|$ **do**
7: $t_{temp} \leftarrow$ find the earliest start time for h on P_i while maintaining the dependencies defined in G and not overlapping with other nodes already on P_i;
8: **if** ($(t_{temp} + Latency(h, P_i)) < t_{rp}$) **then**
9: $P_{target} \leftarrow P_i$;
10: **end if**
11: **end for**
12: $S_{rot} \leftarrow$ Schedule the h onto P_{target};
13: $h \leftarrow$ head.next;
14: **end while**
15: **return** S_{rot}

6.5 Experiments

To evaluate the effectiveness of the proposed algorithms, we conduct experiments on programs from MiBench [82]. The benchmarks used in our experiments include three kinds of representative embedded system applications, multimedia, telecommunication, and security. The result of these experiments show that the proposed algorithms can improve the performance of real applications. Table 6.1 lists the number of tasks and variables in each of the eight benchmarks used in our study. Note that even though each benchmark uses many variables, the access frequency of most variables are very low. We only consider the critical variables, variables of which the access count is not less than 1% of total cycle count. Since the layout of non-critical variables has very limited impact on overall performance, we assign them to SPMs or off-chip memory after all the critical variables have been assigned to SPMs.

To prepare our experiments, we transform C programs of benchmarks to DFGs. We modify a task graph generation tool TGE [211]. We use the modified TGE to divide applications into coarse-gain tasks and statically analyze the programs to extract the critical variables set. In the graph generation process, we guarantee the CCR is no less than 95%; the communication cost can be neglected in scheduling. Second, we use a dynamic analysis tool Valgrind [5] for application profiling. We assume that the on-chip memory is an unlimited sized unified shared memory, and access latency is 10 or 5 clock cycles.

The profiling information includes cycle-accurate execution time and memory reference of tasks, from which we can determine the t_{up} and t_{low} of each task and the VAFM. All the experiments are run on a 2.8 GHZ Pentium 4 PC running Fedora 9 with 1G of DRAM.

Benchmark	**Tasks**	**Vars**	**Benchmark**	**Tasks**	**Vars**
Blowfish enc	12	11	PGP sign	35	8
CRC32	7	6	FFT	25	9
GSM enc	28	12	JPEG	6	15
Lame	16	10	Mad	20	8

TABLE 6.1: Benchmark information.

The "SL" columns represent the schedule length of different approaches. The column $impr_{H:I}$ shows the schedule length improvement of HAFF over IDAS algorithm, and the average improvement is 23.74%. From column $impr_{G:I}$, we can see the significant improvement of GVP algorithm compared to IDAS. The average improvement is 31.91%, the largest improvement is 48.97% for benchmark "JPEG." The last column $impr_{G:H}$ illustrates that the performance of GVP algorithm is better than HAFF 10.40% in average. The critical factor is that GVP has more effective priority metrics for solving the partition in global view than local greedy algorithm HAFF. The results show that the algorithm jointly considering scheduling and variable partitioning outperforms the simple data assignment method.

For the second experiment, all conditions are the same as the first experiment except the remote SPM access latency, which is now 5 clock cycles. Figure 6.8 shows the relative performance of HAFF and GVP algorithms normalized to the IDSA. The lower bars on the graph indicate better performance. The last group of bars compare the average schedule length of these three methods.

The results illustrate that GVP and HAFF also perform well under lower

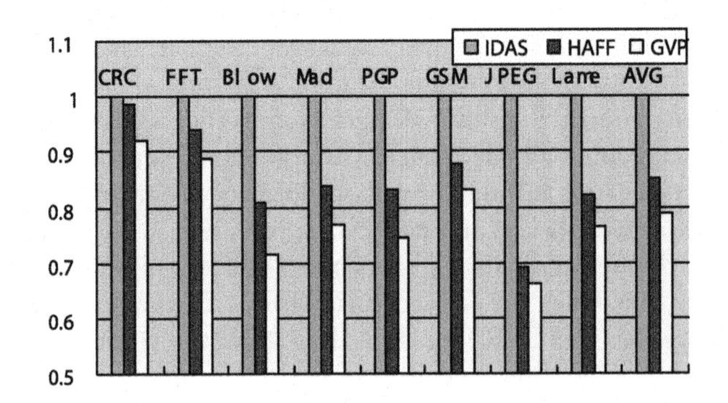

FIGURE 6.8: Compare IDAS, HAFF and GVP with 5-clock cycle latency.

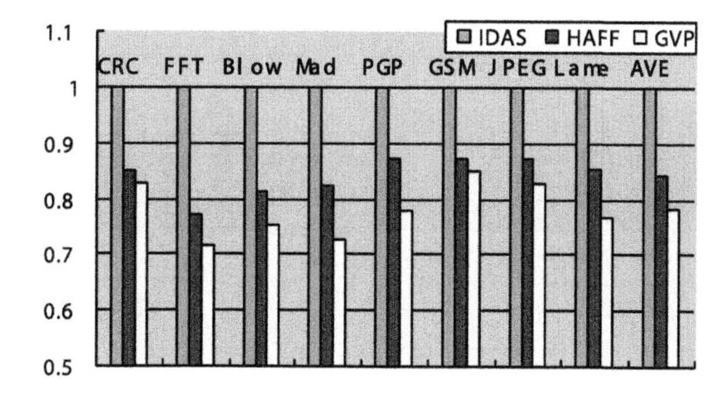

FIGURE 6.9: Rotation scheduling with IDAS, HAFF and GVP.

access latency conditions. In particular, the GVP is very efficient for all the benchmarks.

The first two experiments show variable partitioning is critical for performance improvements in MPSoC with VS-SPM. Our proposed algorithms are good at reducing the number of remote accesses, particularly we can obtain 21.3% average improvement from GVP in low access latency condition. However, in above two experiments, the proposed algorithms are only compared with very simple variable assignment schema (IDAS). The next two experiments will compare them with the optimal variable partition.

The percentage difference between the schedule length of ES-list and GVP-list are shown in "*Diff*". We can see the average schedule length of GVP-list is only 8.74% longer than results of optimal schedule. In some cases, the results generated by GVP-list are very close to the optimal schedule. For example, the schedule length of benchmark "blow" is only 1.4% longer than optimal schedule. For the loop intensive applications, the GVP-RSVP outperforms non-pipeline approach. The average improvement over ES-list is 25.96%. Although the performance of loop pipeline is limited by the inherent parallelism of application, the GVP-RSVP can explore the parallelism as far as possible. For example, the improvement of GVP-RSVP for "CRC32" over ES-list is 45.53%. From Algorithm 6.4.3, we know that rotation is an iterative process, and the variables partitioning may be repeated many times. So exhaustive search variable partitioning cannot be executed in the RVSP algorithm. GVP has very good performance with respect to output schedule length and running time. Particularly, our technique can significantly improve the overall throughput for the applications with high inherent parallelism.

In the fourth experiment, we assume the remote SPM access latency is 5 clock cycles. We compare the schedule length of rotation scheduling with three variable partition algorithms described in experiment one. The results of rotation with HAFF and GVP are normalized to rotation with IDSA shown in Figure 6.9. The average improvement of GVP and HAFF are 21.8% and 15.8%. Another fact indicated by Figure 6.9 is that, in several cases, such as "CRC32" and "JPEG", the results of HAFF approximate to GVP. According

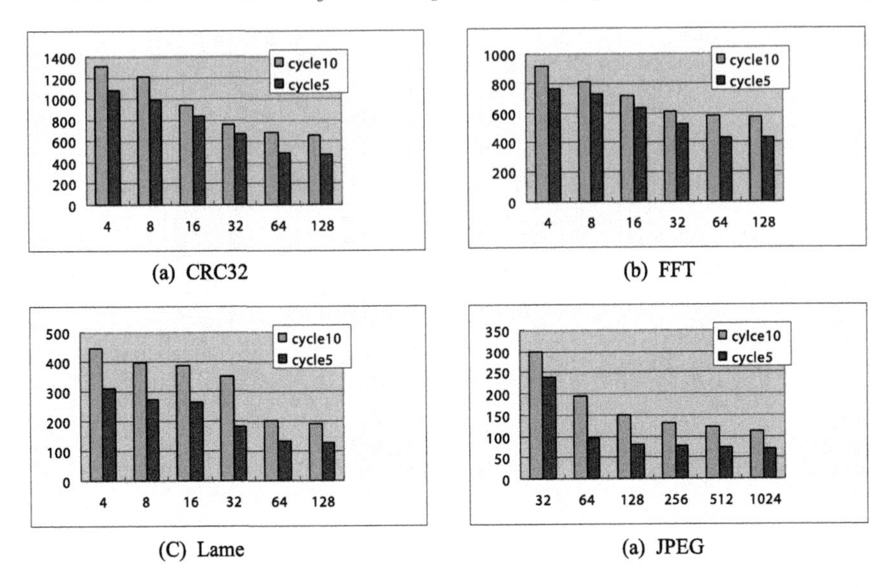

(a) CRC32 (b) FFT

(C) Lame (a) JPEG

FIGURE 6.10: Schedule lengths vary with capacities of SPMs.

to the termination condition of RSVP Equation (6.11), the number of rotation rounds relate to the problem scale. When the problem scale is very large, the low time complexity algorithm HAFF is an acceptable partition algorithm for RSVP.

We also conducted experiments to study the impact of SPMs capability on the schedule length. Figure 6.10 shows the relationship between schedule length and SPMs capability. The schedules are produced by rotation with GVP and HAFF algorithms. The capability of each SPM varies from 4k to 1M and 4 processor cores. The experimental results show that the increasing of SPMs capability does improve the schedule lengths. However, when the capability is larger than a threshold, there is no further improvement from increasing SPMs capability. For example, in Figure 6.10(d), when the SPM size is larger than 64k, the improvement is very little. These experimental results also show that our GVP algorithm outperforms HAFF algorithm for any SPM budget.

6.6 Conclusion

From above sections, we know that the variable partition is very important for MPSoC with VS-SPM. The proposed graph model DFG is efficient for presenting variable access information and loop carried dependencies. In

GVP algorithm, the novel "Weight" matrix estimates the possible variable assignments with global view, which helps GVP to generate a near-optimal variable partition. The performance of HAFF is also good when the remote access latency is low. Both GVP and HAFF are very efficient for large-scale problems. Our loop pipeline technique RSVP is an effective heuristic for loop-intensive application. It exploits the loop parallelism sufficiently with only a few variable repartitioning. The experimental results on selected benchmarks show that our proposed algorithms are promising and they can be applied to different remote access latency conditions and different SPMs capacity.

6.7 Glossary

DRAM: Dynamic Random Access Memory (DRAM) is a type of random access memory that stores each bit of data in a separate capacitor within an integrated circuit.

Integer Linear Programming: A mathematical method for determining a way to achieve the best outcome (such as maximum profit or lowest cost) in a given mathematical model for some list of requirements represented as linear equations.

MPSoC: A system-on-a-chip (SoC) which uses multiple processors (see multicore), usually targeted for embedded applications.

PCM: Phase-change memory is a type of non-volatile computer memory. PCM uses the unique behavior of chalcogenide glass, which can be "switched" between two states, crystalline and amorphous, with the application of heat.

SPM: A small on-chip memory that is controlled by the software, application programs or compiler support.

SRAM: Static Random Access Memory (SRAM) is a type of semiconductor memory. It uses bistable latching circuitry to store each bit.

Chapter 7

Task Scheduling in Multiprocessor to Reduce Peak Temperature

7.1 Introduction

Chip multiprocessor (CMP) has been used in embedded systems, thanks to the tremendous computation requirements in modern embedded processing. Increasing the integration density and achieving higher performance without corresponding increases in frequency are primary goals for microprocessor designers. However the traditional two-dimensional planar CMOS fabrication process are poor at communication latency and integration density. The three-dimensional (3D) CMOS fabrication technology is one of the solutions for faster communication and more functionality on chip. Stacking two or more silicon layers in a CMP, more functional units can be implemented. Meanwhile, the vertical distance is shorter than the horizontal distance in a multi-layer chip [208, 29].

In CMPs, high on-chip temperature impacts circuit reliability, energy consumption, and system cost. Researches show that a 10°C to 15°C increase of operation temperature reduces the lifetime of the chip by half [2]. The increasing temperature causes the leakage current of a chip to increase exponentially. Also, the cooling cost increases significantly which amounts to a considerable portion of the total cost of the computer system. The 3D CMP architecture magnifies the thermal problem. The cross-sectional power density increases linearly with the number of stacked silicon layers, causing a more serious thermal problem.

To mitigate the thermal problem, *Dynamic thermal management* (DTM) techniques such as *dynamic voltage and frequency scaling* (DVFS) have been developed at the architecture level. When the temperature of the processor is higher than a threshold, the DTM can reduce the processor power and control the temperature of the processor. With DTMs, the system performance is degraded inevitably. On the other hand, the operation system level task scheduling mechanism are another way to alleviate the thermal condition of the processor. They either arrange the task execution order in a designated manner, or they migrate "hot" threads across cores to achieve thermal balance. However, most of these thermal-aware task scheduling methods focus on independent tasks or tasks without inter-iteration dependencies. The applications in modern embedded systems often consist of a number of tasks with data dependencies, including inter-iteration dependencies. So it is important to consider the data dependencies in the thermal-aware task scheduling.

In this chapter, we propose a real-time constrained task scheduling algorithm to reduce peak temperature in a 3D CMP. The proposed algorithm is based on the rotation scheduling [41], which optimizes the execution order of dependent tasks in a loop. The main contributions of this chapter include:

- We present an online 3D CMP temperature prediction model.
- We also propose an OS level task scheduling algorithm to reduce the peak temperature. The data dependencies, especially inter-iteration dependencies in the application are well considered in our proposed algorithm.

In Section 7.2, we discuss works related to this topic. In Section 7.3, models for task scheduling in 3D CMPs are presented. A motivational example is given in Section 7.4. We propose our algorithms in Section 7.5, followed by experimental results in Section 7.6. Finally, we give the conclusion in Section 7.7.

7.2 Related Work

Weiser et al. first discussed the problem of task scheduling to reduce the processor energy consumption in [226]. Authors in [136] proposed several schemes to dynamically adjust processor speed with slack reclamation based on the DVS technique. But they only consider the uniprocessor system. Energy-aware task scheduling in parallel system has been studied in the literature recently. Researches in [188, 189] focused on heterogeneous mobile ad hoc grid environments. Authors in those works studied the static resource allocation for the application composed of communicating subtasks in an ad hoc grid. Authors in [206] proposed an energy-aware task scheduling mechanism, EcoMapS. EcoMapS incorporates channel modeling, concurrent task mapping as well as communication and computation scheduling. The proposed scheduling mechanism does not consider the real-time constraint. Authors in [162] proposed two task scheduling algorithms for embedded system with heterogeneous functional units. One of them is optimal and another is near-optimal heuristic. The task execution time information was stochastically modeled. In [167], the authors proposed a loop scheduling algorithm for voltage assignment problem in embedded systems.

In chip design stage, several techniques are implemented for thermal-aware optimization. Authors in [154, 245, 12, 84] propose different thermal-aware 3D floorplanning algorithms. Task allocation and scheduling is another approach to reduce temperature on the chip. Several temperature-aware algorithms are present in [57, 17, 249, 248, 122]. The Adapt3D approach in [57] assigns the upcoming job to the coolest core to achieve thermal balance. The method in [248] is to wrap up aligned cores into super core. Then the hottest job is assigned to the coolest super core. In [17], a thermal management scheme incorporates temperature prediction information and runtime workload characterization to perform efficient thermally aware scheduling. A scheduling scheme based on mathematic analysis is proposed in [249]. Authors in [122] present a slack selection algorithm with a stochastic workload model for thermal-aware dynamic frequency scaling. But none of these approaches considers the inter-iteration dependencies in an application.

7.3 Model and Background

7.3.1 Thermal Model

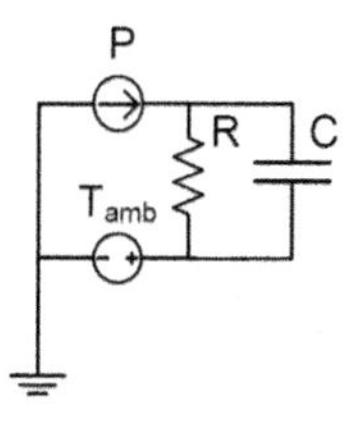

FIGURE 7.1: A Fourier thermal model of a single block.

Fourier heat flow analysis is the standard method of modeling heat conduction for circuit-level and architecture-level IC chip thermal analysis [249]. It is analogous to Georg Simon Ohm's method of modeling the electrical current. A basic Fourier model of heat conduction in a single block on a chip is shown in Figure 7.1. In this model, the power dissipation is similar to the current source and the ambient temperature is analogous to the voltage source. The heat conductance of this block is a linear function of conductivity of its material and its cross-sectional area divided by its length. It is equivalent to electrical conductance. And the heat capacity of this block is analogous to the electrical capacitance. Assuming there is a block on a chip with heat parameters as the ones in Figure 7.1, we have the Fourier heat flow analysis model:

$$C\frac{d(T(t) - T_{amb})}{dt} = P - \frac{T(t) - T_{amb}}{R} \tag{7.1}$$

C is the heat conductance of this block. $T(t)$ is the temperature of that block at time t. T_{amb} is the ambient temperature. P is the power dissipation. And R is the heat resistance. By solving this differential equation, we can get the temperature of that block as follows:

$$T(t) = P \times R + T_{amb} - (P \times R + T_{amb} - T_{init})e^{-t/RC} \tag{7.2}$$

T_{init} is the initial temperature of that block.

Considering there is a task a running on this block, and the corresponding power consumption is P_a, we can predict the temperature of this block by Equation (7.2). Assuming that the execution time of a is t_a, we have the temperature of this block when a is finished:

$$T(t_a) = P_a \times R + T_{amb} - (P_a \times R + T_{amb} - T_{init})e^{-t_a/RC} \tag{7.3}$$

When the execution of task a goes infinite, the temperature of this block reaches a stable state, T_{ss}, which is as follows:

$$T_{ss} = P_a \times R + T_{amb} \tag{7.4}$$

Substituting Equation (7.4) in Equation (7.3), we can get an alternative way of predicting the finish temperature of task a running on that block:

$$T(t_a) = (T_{ss} - T_{init})(1 - e^{-t_a/RC}) + T_{init} \tag{7.5}$$

We can further simplify equation (7.5) as follow:

$$T(t_a) = (T_{ss} - T_{init})(1 - e^{-bt_a}) + T_{init} \tag{7.6}$$

where $b = 1/RC$.

7.3.2 The 3D CMP and the Core Stack

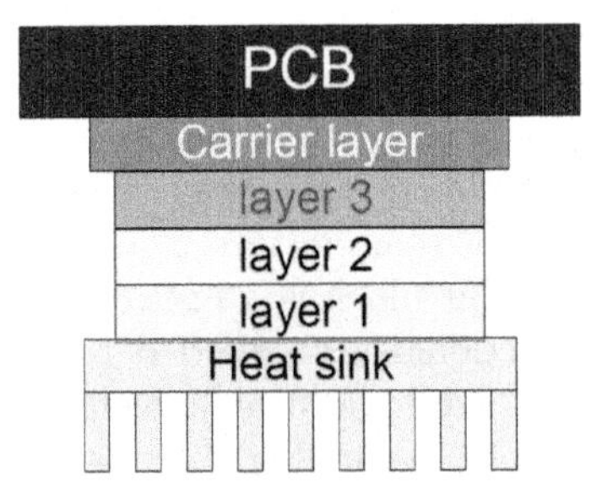

FIGURE 7.2: The cross-sectional view of a 3D chip.

A 3D CMP consists of multiple layers of active silicon. On each layer, there exist one or more processing units, which we call cores. Figure 7.2 shows a basic multi-layer 3D chip structure. A heat sink is attached to the top of the chip to remove the heat off the chip more efficiently. The horizontal lateral heat conductance is approximately 0.4 W/K (i.e., "R_a" in Figure 7.3), much less the conductance between two vertically aligned cores (approximately 6.67 W/K, i.e., "R_2" in Figure 7.3) [249]. The temperatures of vertically aligned cores are highly correlated, relative to the temperatures of horizontally adjacent cores. Therefore, in our online temperature prediction model using our scheduling algorithms, we ignore the horizontal lateral heat conductance. Note that the simulation in our experiment is general enough to consider both the horizontal lateral heat conductances and the vertical ones. We call a set of vertically aligned cores a core stack. Cores in a core stack are highly thermal correlative. High temperature of a core caused by heavy loading will also increase the temperatures of other cores in the core stack. For cores in a core stack, the distances from them to the heat sink are different. Considering a number k of cores in a core stack, where core k is furthest from the heat sink and core 1 is closest to the heat sink, the stable state temperature of the core j ($j \leq k$) can be calculated as,

$$T_{ss}(j) = \sum_{i=1}^{j}(\sum_{l=i}^{k} P_l \times R_i) + T_{amb} \tag{7.7}$$

where P_l is the power consumption of the core l and R_i is the inter-layer thermal conductance between cores i-1 and i (see Figure 7.4).

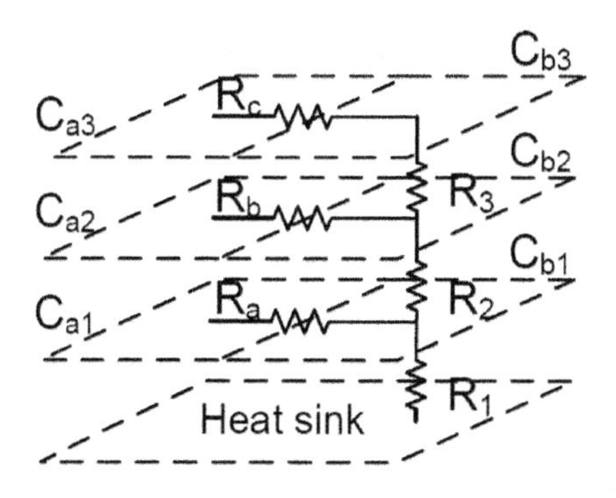

FIGURE 7.3: The Fourier thermal model of a 3D chip.

In order to predict the finish temperature of task a running on core j online, we approximate this finish temperature $T_j(t_a)$ by substituting Equation (7.7) in Equation (7.5) as

$$T_j(t_a) = (\sum_{i=1}^{j}(\sum_{l=i}^{k} P_l \times R_i) + T_{amb} - T_{init\_j}) \times (1 - e^{-t_a/R_jC_j}) + T_{init\_j} \tag{7.8}$$

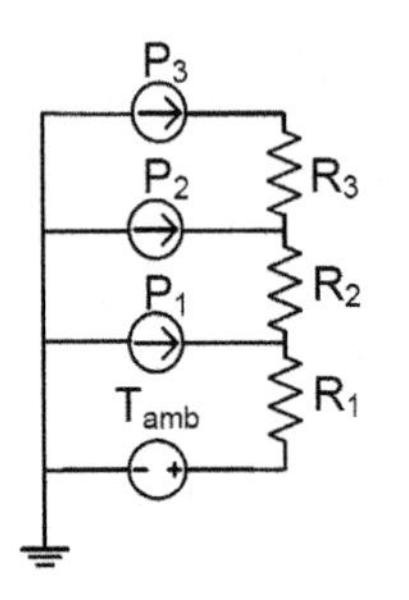

FIGURE 7.4: A simplified Fourier thermal model of a 3D chip.

7.3.3 Application Model

Data-Flow Graph (DFG) is used to model an embedded system application. A DFG typically consists of a set of vertices V, each of which represents a task in the application, and a set of edges E, showing the dependencies among the tasks. The edge set E contains edges e_{ij} for each task $v_i \in V$ that task $v_j \in V$ depends on. The weight of a vertex v_i represents the task type of the task i. Also the weight of an edge e_{ij} means the size of data which is produced by v_i and required by v_j.

We use a cyclic DFG to represent a loop of an application in this chapter. In a cyclic DFG, a delay function $d(e_{ij})$ defines the number of delays for edge e_{ij}. For example, assuming $d(e_{ab}) = 1$ is the delay function of the edge from task a to b, it means the task b in the ith iteration depends on the task a in the $(i-1)^{th}$ iteration. In a cyclic DFG, edges without delay represent the intra-iteration data dependencies, while the edges with delays represent the inter-iteration dependencies. One delay is denoted as a bar. There is a real-time constraint L. It is the deadline of finishing one period of the application. To generate a schedule of tasks in a loop, we use the static *direct acyclic graph* (DAG). A static DAG is a repeated pattern of an execution of the corresponding loop. For a given cyclic DFG, a static DAG can be obtained by removing all edges with delays.

Retiming is a scheduling technique for cyclic DFGs considering inter-iteration dependencies [41]. Retiming can optimize the cycle period of a cyclic DFG by distributing the delays evenly. For a given cyclic DFG G, the retiming function $r(G)$ is a function from the vertices set V to integers. For a vertex u_i of G, $r(u_i)$ defines the number of delays drawn from each of incoming edges of node u_i and pushed to all of the outgoing edges. Let a cyclic DFG G_r be the cyclic DFG retimed by $r(G)$, then for an edge e_{ij}, $d_r(e_{ij}) = d(e_{ij}) + r(v_i) - r(v_j)$ holds, where $d_r(e)$ is the new delay function of edge e_{ij} after retiming and $d(e_{ij})$ is the original delay function.

7.4 Motivational Example

7.4.1 An Example of Task Scheduling in CMP

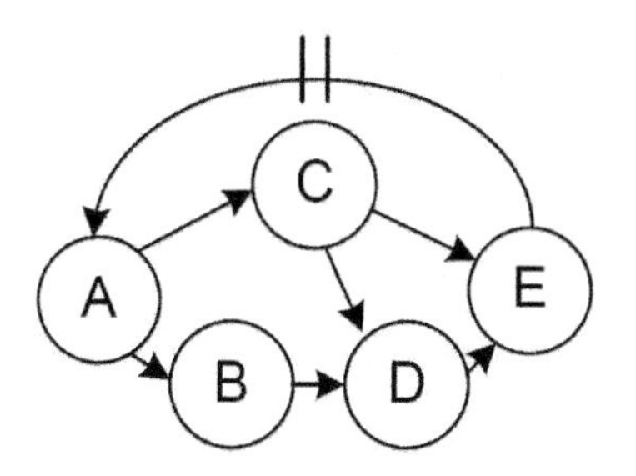

FIGURE 7.5: The DFG of an application.

First we give an example for task scheduling in a multicore chip. We schedule an application (see Figure 7.7) in a two-core embedded system. A DFG representing this application is shown in Figure 7.5. There are two different cores in one layer. The execution times (t) and the stable state temperatures (T_{ss}) of each task in this application running on different cores are shown in Figure 7.6. For simplicity, we provide the stable state temperatures instead of power consumptions in this example, and we assume the value of b (see Equation (7.6)) in each core is the same, 0.025. We also assume the initial temperatures and the ambient temperatures are 50°C.

Task	P1		P2	
	t	Tss	t	Tss
A	95	84	65	92
B	70	80	60	78
C	60	95	65	82
D	80	82	80	70
E	50	75	75	82

FIGURE 7.6: The characteristics of this application.

```
for(i=2;i<N;i++) {
 A[i]=TaskA(E[i-2]);
 B[i]=TaskB(A[i]);
 C[i]=TaskC(A[i]);
 D[i]=TaskD(B[i],C[i]);
 E[i]=TaskE(C[i],D[i]); }
```

FIGURE 7.7: The pseudo code of this application.

7.4.2 List Scheduling Solution

We first generate a schedule by list-scheduling algorithm. Figure 7.8(b) shows a static DAG, which is transformed from the DFG (see Figure 7.8(a)) by removing the delay edge. In the list scheduling, a task assigning order is generated based on the node information in the DAG, and the tasks are assigned to the "coolest" cores in that order. A schedule is generated as Figure 7.9. With the Equation (7.5), we can get the peak temperature of each task as Figure 7.10. Task A has the highest peak temperatures in both first two iterations. In the first iteration, task A starts at the temperature of 50°C and ends at the temperature 80.84°C. And in the second iteration, task A starts immediately after task B of the first iteration finishes, which means it starts at the temperature of 67.89°C. Since it has a higher initial temperature, the peak temperature (82.50°C) in this iteration is higher.

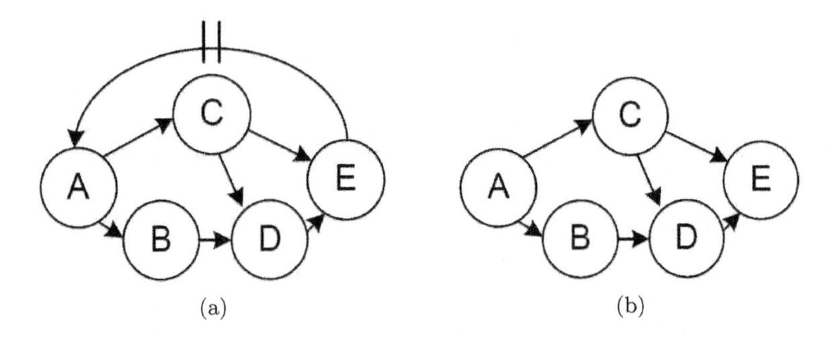

FIGURE 7.8: List scheduling. (a) The DFG. (b) The static DAG.

7.4.3 Improved Solution

Our proposed algorithm uses rotation scheduling to further reduce peak temperature. In Figure 7.11(a), we transform the old DFG into a new DFG by moving a delay from edge e_{EA} to edges e_{AB} and e_{AC}. The new corresponding static DAG is shown in Figure 7.11(b). In this new DAG, there are two parts: node A and the rest nodes. There is no dependency between node A and the

time	P0		P1	
	task	Temperature	task	Temperature
0~95	A	50~80.84	ID	50
95~155	ID	80.84~50.18	B	50~71.75
155~220			C	71.75~79.98
220~300			D	79.98~71.35
300~350	E	50.18~67.89	ID	71.35~56.11

FIGURE 7.9: The schedule generated by list scheduling.

Peak temperature	Task				
	A	B	C	D	E
Iteration 1	80.84	71.75	79.98	71.35	67.89
Iteration 2	82.50	71.88	80.01	71.35	67.89

FIGURE 7.10: The peak temperature (°C) of each task.

rest nodes. The new pseudo code of this new DFG is shown in Figure 7.12, where the operation "A[i+1]=TaskA(E[i-1]);" can be placed anywhere in the loop, due to its independence.

In this case, we can first assign the dependent nodes (B to E) to cores with the same policy used in list scheduling. Tasks B, C and D are assigned to core P1 at the time slot of [0, 205]. And task E is scheduled to run on core P0 at [205, 255]. In this partial schedule, we find out that there are three time slots at which we can schedule task A. One is the idle gap of core P0 at [0, 205], another is the time slot after task E is done (time 255) on P0, and the last one is time slot after task D (time 205) on P1. Because the peak temperature of task A is lowest when running in the idle gap of core P0 at [0, 205], this time slot is selected. Since task A runs after the last iteration of task E, the longer the idle gap between them, the cooler initial temperature task A starts at. So we schedule task A to start at time 110. A schedule is shown in Figure 7.13. In this schedule, the peak temperature is 81°C when task A is running in the second iteration (see Figure 7.14). Our approach reduces the peak temperature by 1.5°C. In addition, the total execution time in one iteration is only 255, while the total execution time generated by list scheduling is 350.

In the next section, we will discuss our thermal-aware task scheduling algorithm which deeply explores the solution space to find the good schedule meeting the real-time constraints.

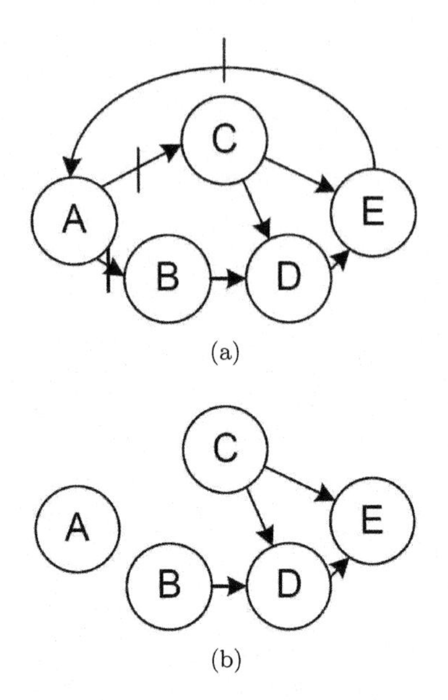

FIGURE 7.11: Rotation scheduling. (a) The retimed DFG. (b) The new static DAG.

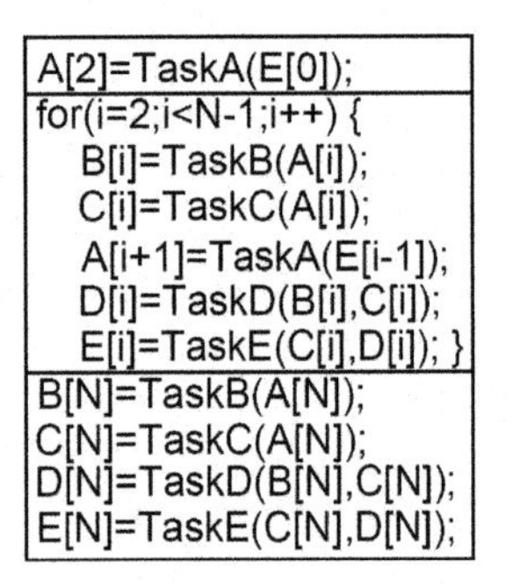

```
A[2]=TaskA(E[0]);
for(i=2;i<N-1;i++) {
   B[i]=TaskB(A[i]);
   C[i]=TaskC(A[i]);
   A[i+1]=TaskA(E[i-1]);
   D[i]=TaskD(B[i],C[i]);
   E[i]=TaskE(C[i],D[i]); }
B[N]=TaskB(A[N]);
C[N]=TaskC(A[N]);
D[N]=TaskD(B[N],C[N]);
E[N]=TaskE(C[N],D[N]);
```

FIGURE 7.12: The retimed pseudo code.

time	P0		time	P1	
	task	Temperature		task	Temperature
0~110	ID	50	0~60	B	50~71.75
			60~125	C	71.75~79.98
110~205	A	50~80.04	125~205	D	79.98~71.35
205~255	E	80.04~67.89	205~255	ID	71.35~56.11

FIGURE 7.13: The schedule generated by our proposed algorithm.

Peak temperature	Task				
	A	B	C	D	E
Iteration 1	80.84	71.75	79.98	71.35	76.67
Iteration 2	81.00	73.12	80.25	71.39	76.72

FIGURE 7.14: The peak temperature (°C) of each task.

7.5 Thermal-Aware Task Scheduling Algorithm

In this section, we propose an algorithm, TARS (*Thermal-Aware Rotation Scheduling*), to solve *the minimum peak temperature without violating real-time constraints problem*. By repeatedly rotating down delays in DFG, more flexible static DAGs are generated. For each static DAG, greedy heuristic is used to generate a schedule with minimum peak temperature. Then the best schedule is selected among the schedules generated previously.

7.5.1 The TARS Algorithm

Algorithm 7.5.1 The TARS Algorithm

Require: A DFG, the rotation times R
Ensure: A schedule S, the retiming function r

1: rot_cnt ← 0 /*Rotation counter.*/
2: Initial S_{min}, r_{min}, PT_{min}, r_{cur} /*The optimal schedule, the according retiming function, the according peak temperature, and the current retiming function*/
3: **while** rot_cnt < R **do**
4: Transform the current DFG to a static DAG.
5: Schedule tasks with dependencies, using the PTMM algorithm.
6: Schedule independent tasks, using the MPTSS algorithm.
7: Get the peak temperature PT_{cur} of the current schedule.
8: **if** $PT_{cur} < PT_{min}$ and S_{cur} meet the real-time constraint **then**
9: $S_{min} \leftarrow S_{cur}$, $r_{min} \leftarrow r_{cur}$, $PT_{min} \leftarrow PT_{cur}$
10: **end if**
11: Use RS algorithm to get a new retiming function r_{cur}
12: Get the new DFG based on r_{cur}
13: $R \leftarrow R + 1$
14: **end while**
15: Output the S_{min}, r_{cur}

In the TARS algorithm shown in Algorithm 7.5.1, we will try to rotate the original DFG by R times. In each rotation, we get the static DAG from the rotated DFG by deleting the delayed edges in DFG. A static DAG usually consists of two kinds of tasks. One kind of task is the task with dependencies, like the tasks B, C, D, and E in Figure 7.11(b). The other kind of tasks is the independent tasks, like the task A in Figure 7.11(b). The independent tasks do not have any intra-iteration relation with other tasks. We first assign tasks with dependencies by the PTMM algorithm.

7.5.2 The PTMM Algorithm

The *Peak Temperature Min-Min* (PTMM) algorithm is designed to schedule the tasks with dependencies. Min-min is a popular greedy algorithm [94]. The original min-min algorithm does not consider the dependencies among tasks. So in the min-min baseline algorithm used in this chapter, we need to update the assignable task set in every step to maintain the task dependencies. We define *the assignable task* as the unassigned task whose predecessors all have been assigned. Since the temperatures of the cores in a core stack are highly correlated in 3D CMP, we need to schedule tasks with consideration of vertical thermal impacts. When we consider assigning a task T_i to core C_j, we calculate the peak temperatures of cores in the core stack of C_j during the T_i running on C_j, based on Equation (7.8). Let $T_{max}(i,j)$ be the maximum value of the peak temperatures in the core stack. So when we decide the assigning of T_i, we calculate all the $T_{max}(i,j),\ for\ j = every\ core$. Due to the fact that the available times and the power characteristics of different cores in the same core stack may not be identical, the peak temperatures of the given core stack may be various when assigning the same task to different cores of this core stack, respectively. Let C_{min} be the core with minimum $T_{max}(i,j)$. In each step in PTMM, we first find all the assignable tasks. Then we will form a pair $<T_i, C_{min}>$ for every assignable task. Only the $<T_i, C_{min}>$ pair which gives the minimum $T_{max}(i,j)$ will be assigned accordingly. And we also schedule the start execution time of T_i as the time when the predecessors of T_i are done and core C_{min} is ready. The PTMM is shown as Algorithm 7.5.2.

7.5.3 The MPTSS Algorithm

Using the PTMM algorithm, we can get a partial schedule, in which the tasks with dependencies are assigned and scheduled. We need to further assign the independent tasks in the static DAG. Since the independent tasks do not have any intra-iteration relations with others, they can be scheduled to any possible time slots of the cores. In the *Minimum Peak Temperature Slot Selection* (MPTSS) algorithm, we assign the independent tasks in the decreasing order of their power consumption. Tasks with larger power consumption likely generate higher temperatures. The higher assigning orders these tasks have, the better fitting cores these tasks will be assigned to, and probably the lower resulting peak temperature of the final schedule.

Before we assign an independent task T_i, we first find all the idle slots

Algorithm 7.5.2 The PTMM Algorithm

Require: A static DAG G, m different cores, EP matrix
Ensure: A schedule generated by PTMM

1: Form a set of assignable tasks P
2: **while** P is not empty **do**
3:     **for** $t =$ every task in P **do**
4:         **for** $j = 1$ to m **do**
5:             Calculate the peak temperatures of cores in the core stack of C_j, assuming t is running on C_j. And find the minimum peak temperature $T_{max}(t, j)$
6:         **end for**
7:         Find the core $C_{min}(t)$ giving the minimum peak temperature $T_{max}(t, j)$
8:         Form a task-core pair as $<t, C_{min}(t)>$
9:     **end for**
10:     Choose the task-core pair $<t_{min}, C_{min}(t_{min})>$ which gives the minimum $T_{max}(t, C_{min}(t))$
11:     Assign task t_{min} to core $C_{min}(t_{min})$
12:     Schedule the start time of t_{min} as the time when all the predecessors of t_{min} are finished and $C_{min}(t_{min})$ is ready.
13:     Update the assignable task set P
14:     Update time slot table of core $C_{min}(t_{min})$ and the expected finish time of t_{min}
15: **end while**

among all cores, forming a time slot set TS. The time slots which are not long enough for the execution of T_i will be removed from TS. Then we calculate the peak temperature of the according core stack $T_{max}(i, j)$, which is defined in the PTMM algorithm, for every time slot. One problem arises here: since the remaining time slots are long enough for the execution of T_i, we need to decide when to start the execution.

We use two different schemes here. The first one is the *As Early As Possible* (AEAP), which means the task T_i should be scheduled to start at the beginning of that time slot. The other one is *As Late As Possible* (ALAP), which means we should schedule the start execution time of the task T_i at a certain time so that T_i will finish at the end of the time slot. These two schemes result in different impacts on peak temperature. We present our MPTSS algorithm in Algorithm 7.5.3.

7.5.4 The RS Algorithm

At the end of each iteration of the TARS algorithm, we create a new DFG by rotating the current DFG. First, we need to form a set of rotation tasks. If a task is the first task scheduled on a core and there is at least one delay in

Algorithm 7.5.3 The MPTSS Algorithm

Require: A partial schedule generated by PTMM, a set of independent tasks, m different cores, EP matrix
Ensure: A schedule generated by MPTSS
1: List independent tasks in a list P in the decreasing order of their power consumption
2: **while** The list P is not empty **do**
3: $t = top(P)$
4: Collect all the time slots which is long enough for t across all cores, form a time slot set TS.
5: **for** Every time slot ts_i in TS **do**
6: $j \leftarrow$ the according core of ts_i
7: Find the task t_{next} which is scheduled to start right after ts_i on the core C_j.
8: **if** $Power(t) < Power(t_{next})$ **then**
9: Find the start time with the AEAP scheme
10: **else**
11: Find the start time with the ALAP scheme
12: **end if**
13: Get the $T_{max}(t, j)$ /*similar to the one in PTMM*/
14: **end for**
15: Find the time slot ts_{min} giving the minimum peak temperature $T_{max}(t, j)$
16: Assign task t to core C_{min} /*C_{min} is the core of time slot ts_{min}*/
17: Schedule the start time of t in time slot ts_{min} based on the scheme selected in the if statement (line 8).
18: Remove t from P
19: Update time slot table of core C_{min}
20: **end while**

each of its incoming edge, this task is a rotation task. The *Rotation Scheduling* (RS) algorithm is shown in Algorithm 7.5.4.

7.6 Experimental Results

This section presents the experimental results of our algorithm. We develop our experiments as follows: we first use a precise microprocessor simulator, Wattch 1.0.2 [35] to get the execution and power characteristic of a set of benchmarks. Then we generate a number of random DFGs consisting of this set of benchmarks. Task schedules and power traces are created by our

Algorithm 7.5.4 The RS Algorithm

Require: An input DFG D_{in} and a schedule S based on D_{in}, a retiming function r

Ensure: An output DFG D_{out} generated by rotation scheduling, a new retiming function r_{new}

1: Form the set of rotation tasks RT based on D_{in} and S
2: **for** Every task t_i in RT **do**
3: Reduce one delay from every incoming edge of task t_i in D_{in}
4: Increase one delay from every outgoing edge of task t_i in D_{in}
5: $r(t_i) \leftarrow r(t_i) + 1$
6: **end for**
7: $D_{out} \leftarrow D_{in}$ and $r_{new} \leftarrow r$

algorithm. We input these schedules and power traces into a thermal analysis simulator, called Hotspot 4.1 [191]. Finally, we evaluate our algorithm with the comprehensive thermal analysis generated by Hotspot 4.1. All experiments are conducted on Linux machine equipped with an Intel Core 2 Duo E8400 CPU and 3GB of RAM.

7.6.1 Experiment Setup

The 3D CMP architecture simulated in our experiments is a two-layer front-to-back architecture. There are eight Alpha 21264 (EV6) microprocessor cores in each layer with configuration as Table 7.1. We use per core DFVS in our simulation with three DVFS levels (3.88GHz, 4.5GHz, and 5 GHz) configured based on the parameters of Alpha 21264 [106].

Processor core	Alpha 21264
Core technology	65nm
Nominal frequency	5GHz
L1 data cache	64K, 2-way
L1 instruction cache	64K, 2-way
L2 cache	2M

TABLE 7.1: Configuration of Alpha cores.

We choose the SPEC CPU 2000 benchmark suite in our experiment. The execution time and the power consumption of each benchmark on Alpha core are tested through the Wattch 1.0.2 simulator with the above configuration. For each benchmark, we run it under those three DVFS levels via out-of-order mode to get the task characteristic of this benchmark. We generate 10 random DFG based applications. The tasks in these applications are randomly selected from the SPEC2000 benchmarks. For each application, we set the real-time constraint TC (i.e., deadline) as follows:

$$TC = \frac{\sum_{i=1}^{N} t_i}{P} \times c \tag{7.9}$$

where N is the number of tasks in this application, t_i is the execution of time of task i under the highest frequency, P is the total number of cores, i.e., 16 in our simulation, and c is a constant which is set to 5.

The thermal simulation is conducted in the Hotspot 4.1 simulator by using the power consumption traces created by our program. In the Hotspot 4.1 simulator, the lateral and vertical thermal interactions among adjacent core are all carefully considered and modeled. As we mentioned above, the architecture model used in the Hotspot simulator is a two-layer architecture, in which the thickness of the top layer (the one far from the heat sink) is 50μm, and the thickness of the bottom (the one close to the heat sink) is 300μm. There is a thermal interface material (TIM) layer between these two layers. The core size is 4mm $\times$ 8mm. Some other parameters are listed in Table 7.2.

Layer	Conductivity	Capacitance per unit volume
Silicon	$100\ W/(m \cdot K)$	$1.75 \times 10^6\ J/(m^3 \cdot K)$
TIM	$4\ W/(m \cdot K)$	$4.0 \times 10^6\ J/(m^3 \cdot K)$
Copper	$400\ W/(m \cdot K)$	$3.55 \times 10^6\ J/(m^3 \cdot K)$

TABLE 7.2: Thermal parameter for Hotspot.

As our algorithm is to reduce the peak temperature in 3D CMP architectures, we show the average peak temperature of all 16 cores over 10 applications in Figure 7.15. By comparing the result of list scheduling, we find that our algorithm can reduce the peak temperatures by up to 10 °C. For the cores in the top layer (core 1 to 8), the peak temperatures are consistently higher than the ones in the bottom layer (core 9 to 16). This result is aligned to our online thermal prediction model. The peak temperatures of top layer cores is around 83 °C when using the task schedule from our TARS algorithm and is about 90 °C using list scheduling.

Larger improvements are made in the top layer cores. The reason is that in our proposed algorithm, more effort is made in reducing the temperature of the hottest core, which is usually located in the top layer. Even though the improvements for cores in the bottom layer are not as significant as the ones in the top layer, lower peak temperatures are achieved, due to the more flexible execution order explored in our algorithm and less impact from the aligned cores on the top layer. The reducing of peak temperature in the bottom layer is about 2.5°C.

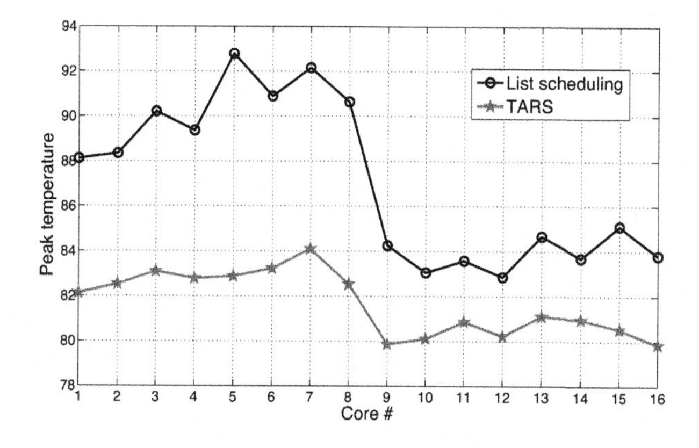

FIGURE 7.15: Core peak temperatures comparison.

7.7 Conclusion

An online 3D CMP temperature prediction model is presented in this chapter. We also propose our real-time constrained task scheduling algorithm to reduce peak temperature in a 3D CMP. By considering the frequencies assignment and the inter-iteration data dependencies collaboratively, our proposed TARS algorithm can significantly reduce the peak temperature on chip. Our simulation results show that our TARS algorithm can reduce peak temperature by 10°C.

7.8 Glossary

3D CMOS: More than one layer of cores is integrated in a single chip.

Alpha 21264: The Alpha 21264 was a Digital Equipment Corporation RISC microprocessor introduced in October 1996.

CMP: Chip multiprocessor. A single integrated circuit die in which the cores are integrated.

DTM: Dynamic thermal management. A trigger mechanism is provided to respond with a throttling of activity in order to guarantee a reliable operation of the device in case of a thermal emergency.

DVFS: Dynamic voltage and frequency scaling is a power management technique in computer architecture, where the voltage and frequency used in a component is increased or decreased, depending upon circumstances.

Chapter 8

Networked Sensing and Monitoring Systems

8.1 Introduction

Sensor networks have wide applications in areas such as health care, military, environmental monitoring, infrastructure security, manufacturing automation, collecting information in disaster prone areas, and surveillance applications [95, 7, 200, 110, 46]. For many applications, such as national security and health care, sensor networks are the backbone of the structures. In this context, a key problem that ought to be tackled is to design embedded software and hardware architectures that can effectively operate in continuously changing, hard-to-predict conditions [101]. In fact, the vision is that sensor networks will offer ubiquitous interfacing between the physical environment and centralized databases and computing facilities [227, 129, 172]. Efficient interfacing has to be provided over long periods of time and for a variety of environment conditions, like moving objects, temperature, weather, available energy resources, and so on.

In many sensor networks, the processors consume a significant amount of energy, memory, buffer size, and time in highly computation-intensive applications [109, 166]. However, sensor node can only be equipped with a limited power source (≤ 0.5 Ah, 1.2 V) [231, 25, 37, 67]. In some application scenarios, replenishment of power resources might be impossible. Therefore, power consumption has become a major hurdle in the design of next generation portable, scalable, and sophisticated sensor networks [63, 62, 192, 237]. In computation-intensive applications, an efficient scheduling scheme can help reduce the energy consumption while still satisfying the performance constraints [101, 97, 183, 19, 65]. This chapter focuses on reducing the total energy of sensor applications on architectures with task scheduling and model assignment.

This chapter presents assignment and optimization algorithms which operate in probabilistic environments to solve the MAP problem. In the MAP problem, we model the execution time of a task as a random variable [91, 207, 247, 92]. For heterogeneous systems [19], each node has a different energy consumption rate, which relates to area, size, reliability, etc.

[64, 85, 97, 120]. Faster one has higher energy consumption while slower one has lower consumption. This chapter shows how to assign a proper mode to each node of a *Probability Data Flow Graph* (PDFG) such that the total energy consumption is minimized while the timing constraint is satisfied with a guaranteed confidence probability. With confidence probability P, we can guarantee that the total execution time of the DFG is less than or equal to the timing constraint with a probability that is greater than or equal to P.

We use adaptive approach for online customization of embedded architectures that function in non-stationary environments [101]. We design an algorithm, TSEO (*Task Scheduling and Energy-aware Online*) algorithm, to minimize energy consumptions while satisfying timing constraints with guaranteed probability. The obtained voltages will affect the adaptation thresholds of control policies.

The experimental results show that TSEO achieves a significant reduction in total energy consumption on average. For example, with 10 sensors, compared with Method 1, TSEO shows an average 38.1% reduction in total energy consumption while satisfying the timing constraint with probability 0.90. Our algorithm has several advantages: First, our algorithm explores the larger solution space for total energy consumption with soft real-time. Second, our TSEO algorithm combined both soft real-time and multi-sensor scheduling, and can significantly reduce total energy consumption while satisfying timing constraints with guaranteed probability. It is efficient and provides an overview of all possible variations of minimum energy consumptions compared with the worst-case scenario generated by Method 1 and Method 2.

Our contributions are listed as the following:

- We propose two highly efficient algorithms to solve task-sensor scheduling problems for sensor network applications. By using multiple sensors scheduling, our algorithms improve both the performance and energy-minimization of sensor networks.

- Our algorithm *MAP_Opt* gives the optimal solution and achieves more significant energy saving than *MAP_CP* algorithm.

- Our algorithm not only is optimal, but also provides more choices of smaller total energy consumption with guaranteed confidence probabilities satisfying timing constraints. In many situations, algorithm *MAP_CP* cannot find a solution, while ours can find satisfied results.

- Our algorithm is practical and efficient.

In the next section, we introduce necessary background, including basic definition and models. The algorithm is discussed in Section 8.3. We show our experimental results in Section 8.4.

8.2 Models

In this section, we introduce some basic concepts and models which will be used in the later sections. First, the heterogeneous FU schedule problem is given. Next, virtual sensor network is introduced. An example is given throughout the whole section.

8.2.1 Heterogeneous FU Schedule Problem

We define the heterogeneous sensors scheduling problem as follows: given a heterogeneous system with m sensors, $F = F_1, \cdots, F_i, \cdots, F_m$, a DAG $G = \langle V, E_d \rangle$ where $V = \langle u_1, \cdots, u_i, \cdots, u_N \rangle$ is a set of task nodes, with each node representing a task, $E_d \subseteq V \times V$ is a set of edges representing dependencies relations among nodes in V. $T(u_k) = t_1(k), \cdots, t_i(k), \cdots, t_m(k)$, where $t_i(k)$ denoted the computation time of u_k on F_i, and a time constraint L, we can find a task schedule for G such that the failure rate is minimized within L.

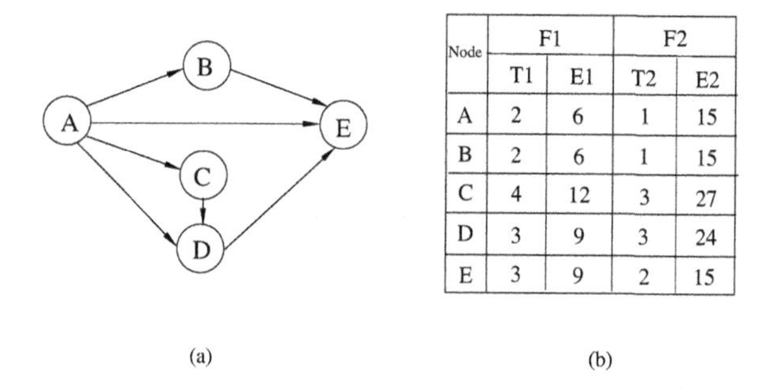

Node	F1		F2	
	T1	E1	T2	E2
A	2	6	1	15
B	2	6	1	15
C	4	12	3	27
D	3	9	3	24
E	3	9	2	15

FIGURE 8.1: (a) A DAG. (b) Execution times and failure rate of FUs.

An example is shown in Figure 8.1 and Figure 8.2. Assume there is a heterogeneous system that consists of two heterogeneous sensors, F1 and F2. An exemplary DAG is shown in Figure 8.1(a). The given time constraint for the DAG to be executed is 10 time units. The execution time and failure rate of each node for different sensors are shown in Figure 8.1(b). In Figure 8.1(b), T_i denotes the execution time and E_i denotes the first part of the failure rate E_{ij}, which is the failure rate for u_i to be scheduled on F_j. Two schedules for the DAG are shown in Figure 8.2. In Schedule 1, the schedule length is 10 time units and its system failure rate is 63. In Schedule 2, the schedule length is 10 time units and its system failure rate is 57. Both schedules satisfy the time constraint while the latter has a lower failure rate. This example shows that different task schedules will produce different system failure rates.

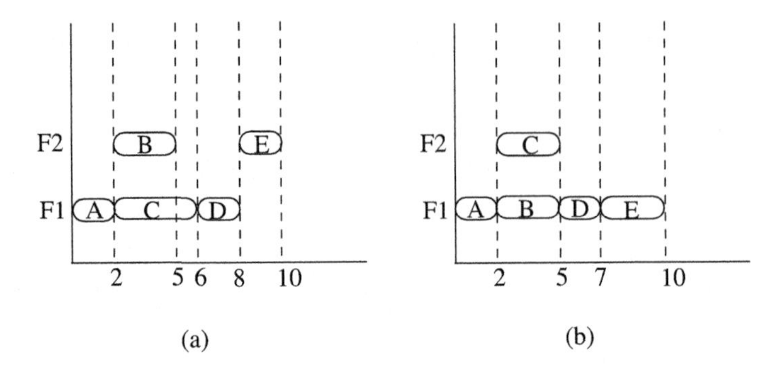

FIGURE 8.2: (a) Schedule 1. (b) Schedule 2.

8.2.2 The Virtual Sensor Network Model

Sensors usually work in highly changing environments. The work load and execution time of each node will change dynamically. For example, consider a camera-based sensor, which is tracking a moving object, such as a person or vehicle [227, 101]. We modeled the system dynamics with discrete events formulated over a fixed time interval or time window used to obtain the future performance requirements of the system.

After, we assign tasks to different sensors. Based on the dependence relationships, the new task graph becomes a virtual sensor network. In our model, under the same mode (M) of a sensor, the execution time (T) of a task is a random variable, which is usually due to condition instructions or operations that could have different execution times for different inputs. The energy consumption (E) depends on the mode M. Under different modes, a task has different energy consumptions. The execution time of a node in active mode is less than that in vulnerable mode, and they both are less than the execution time in sleep mode. The relations of energy consumption are just the reverse. This chapter shows how to assign a proper mode to each node of a *Probabilistic Data-Flow Graph* (PDFG) such that the total energy consumption is minimized while satisfying the timing constraint with a guaranteed confidence probability.

Definition 8.1 We define the *MAP (Mode Assignment with Probability)* problem as follows: Given R different modes: M_1,M_2,$\cdots$,M_R, a PDFG $G = \langle U, ED \rangle$ with $T_{M_j}(v)$, $P_{M_j}(v)$, and $E_{M_j}(u)$ for each node $u \in U$ executed on each mode M_j, a timing constraint L and a confidence probability P, find the mode for each node in assignment A that gives the *minimum total energy consumption E with confidence probability P under timing constraint L.*

For the adaptive architecture, the working procedures are as follows. During look ahead for the next time interval, the work load and corresponding execution time of each node is estimated. In each time interval, we find the best mode assignment to minimize total energy consumption while satisfying timing constraints with guaranteed probabilities. After finding the best assignment, the controller accordingly makes changes to the pool of hardware with the updated policy. The time interval is a design parameter, and will have to be decided by the designer based on empirical data obtained from simulations of the particular application.

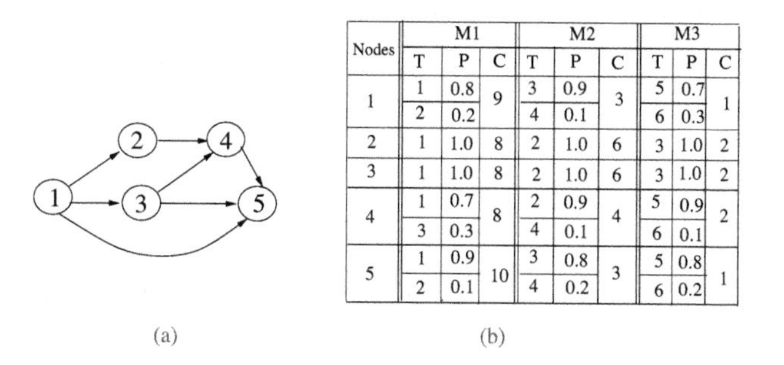

Nodes	M1			M2			M3		
	T	P	C	T	P	C	T	P	C
1	1	0.8	9	3	0.9	3	5	0.7	1
	2	0.2		4	0.1		6	0.3	
2	1	1.0	8	2	1.0	6	3	1.0	2
3	1	1.0	8	2	1.0	6	3	1.0	2
4	1	0.7	8	2	0.9	4	5	0.9	2
	3	0.3		4	0.1		6	0.1	
5	1	0.9	10	3	0.8	3	5	0.8	1
	2	0.1		4	0.2		6	0.2	

FIGURE 8.3: (a) A sensor network topology. (b) The times, probabilities, and energy consumptions of its nodes in different modes.

An exemplary PDFG is shown in Figure 8.3(a), which is transferred from the task-sensor scheduling graph. Each node can select one of the three different modes: M_1 (active), M_2 (vulnerable), and M_3 (sleep). The execution times (T), corresponding probabilities (P), and energy consumption (E) of each node under different modes are shown in Figure 8.3(b). The input DAG (*Directed Acyclic Graph*) has five nodes. Node 1 is a multi-child node, which has three children: 2, 3, and 5. Node 5 is a multi-parent node, and has three parents: 1, 3, and 4. The execution time T of each node is modeled as a random variable. For example, when choosing M_1, node 1 will be finished in 1 time unit with probability 0.8 and will be finished in 2 time units with probability 0.2. Node 1 is the source and node 5 is the destination or the drain.

In sensor network application, a real-time system does not always have a hard deadline time. The execution time can be smaller than the hard deadline time with certain probabilities. So the hard deadline time is the worst-case of the varied smaller time cases. If we consider these time variations, we can achieve a better minimum energy consumption with satisfying confidence probabilities under timing constraints.

For Figure 8.3, the minimum total energy consumptions with computed confidence probabilities under the timing constraint are shown in Table 8.1. The results are generated by our algorithm, *MAP_Opt*. The entries with probability that is equal to 1 (see the entries in boldface) actually give the results to the hard real-time problem which shows the worst-case scenario of the MAP problem. For each row of the table, the E in each (P, E) pair gives the minimum total energy consumption with confidence probability P under timing

T	(P , E)	(P , E)	(P , E)	(P , E)	(P , E)
4	0.50, 43				
5	0.65, 39				
6	0.65, 35	0.81, 39			
7	0.65, 27	0.73, 33	0.81, 35	0.90, 39	
8	0.81, 27	0.90, 35	**1.00, 43**		
9	0.58, 20	0.73, 21	0.81, 27	0.90, 32	**1.00, 39**
10	0.72, 20	0.81, 21	0.90, 28	**1.00, 36**	
11	0.65, 14	0.90, 20	**1.00, 32**		
12	0.81, 14	0.90, 20	**1.00, 28**		
13	0.65, 12	0.90, 14	**1.00, 20**		
14	0.81, 12	0.90, 14	**1.00, 20**		
15	0.50, 10	0.90, 12	**1.00, 14**		
16	0.72, 10	0.90, 12	**1.00, 14**		
17	0.90, 10	**1.00, 12**			
18	0.50, 8	0.90, 10	**1.00, 12**		
19	0.72, 8	**1.00, 10**			
20	0.90, 8	**1.00, 10**			
21	**1.00, 8**				

TABLE 8.1: Minimum total energy consumptions with computed confidence probabilities under various timing constraints.

constraint j. For example, using our algorithm, at timing constraint 12, we can get (0.81, 14) pair. The assignments are shown in Table 8.2. We change the mode of nodes 2 and 3 to be M_2. Hence, we find the way to achieve minimum total energy consumption 14 with probability 0.81 satisfying timing constraint 12. While using the heuristic algorithm *MAP_CP* [183], the total energy consumption obtained is 28. Assignment $A(u)$ represents the voltage selection of each node u. There is 50% energy saving comparing with *MAP_CP*.

8.3 The Algorithms

8.3.1 Task-Sensor Scheduling

In this subsection, two algorithms, *Sensor_Sch1* and *Sensor_Sch2* are designed to solve the Task-Sensor Scheduling problem, i.e., how to obtain the minimum total system energy consumption without sacrificing performance, based on multiple sensor scheduling.

Scheduling problems with time and resource constraints are well-known to be NP complete. We are going to solve a scheduling problem with time and resource constraints in a heterogeneous system and minimize the total energy consumption of the system at the same time. Therefore, our problem is also NP complete. In this section, two novel algorithms, *Sensor_Sch1* and *Sensor_Sch2*, are developed to solve this problem. *Sensor_Sch1* uses a bipartite matching strategy based on ALAP scheduling and *Sensor_Sch2* uses a pro-

		id	T	M	Prob.	E.
Ours	$A(u)$	1	3	M_2	0.90	3
		2	3	M_3	1.00	2
		3	3	M_3	1.00	2
		4	2	M_2	0.90	4
		5	4	M_2	1.00	3
	Total		12		0.81	14
MAP_CP	$A(u)$	1	2	M_1	1.00	9
		2	2	M_2	1.00	6
		3	2	M_2	1.00	6
		4	4	M_2	1.00	4
		5	4	M_2	1.00	3
	Total		12		1.00	28

TABLE 8.2: The assignments of two algorithms *MAP_Opt* and *MAP_CP* with timing constraint 12.

gressive relaxation strategy based on ALAP scheduling. Some symbols used in our algorithms are listed in Table 8.3.

Symbol	**Meaning**
Min	Minimum energy consumption for the current node
EST_i	Earliest starting time for node i
LST_i	Latest starting time for node i
FT_i	Finish time of node i
SL_j	Schedule length for $sensor_j$
X_i	The sensor that node i is scheduled on
E_{total}	Overall energy consumption for the system

TABLE 8.3: Symbols in the sensor scheduling algorithms.

8.3.1.1 Sensor Scheduling with Bipartite Matching

Sensor_Sch1 is designed to use the bipartite matching strategy to schedule tasks. The idea is: first, use the ALAP scheduling, which minimizes the schedule length, to get the latest starting time for every node. Then construct a bipartite matching graph with the nodes in the ready list in one set and all sensors in the other set. Then reschedule nodes based on the minimum energy consumption bipartite matching. This algorithm is shown in Algorithm 8.3.1. V_1 and V_2 in the algorithm represent the two sets used in the bipartite matching.

This algorithm first schedules all nodes using the ALAP scheduling. Based on the schedule, we use the bipartite matching to schedule nodes. For each sensor, among those nodes not marked by matching, the node with the earli-

Algorithm 8.3.1 Sensor_Sch1 Algorithm

Require: a DAG $G = \langle V, E_d \rangle$, a set of sensors, a timing constraint L
Ensure: A task scheduling with energy consumption minimization

```
1: for all u_i ∈ V; do
2:    EST_i ← starting time of node i in ASAP;
3:    LST_i ← starting time of node i in ALAP;
4:    FT_i ← Finish time of node i in ALAP;
5: end for
6: for all F_j ∈ F do
7:    SL_j ← 0;
8: end for
9: E_total ← 0;
10: while ∃ nodes not marked do
11:    for all F_j ∈ F do
12:       if the first node in F_j has no dependency constraint then
13:          put it into V_1;
14:       end if
15:       put F_j into V_2;
16:    end for
17:    construct a weighted bipartite matching graph G_BM = ⟨V_BM, E_BM⟩; V_BM =
       V_1 ∪ V_2;
18:    for all u_i ∈ V_1 do
19:       for all F_j ∈ V_2 do
20:          compute E_ij;
21:          add an edge e_ij between u_i and F_j into E_BM;
22:          if SL_j + t_j(i)+ data migration delay ≤ FT_i then
23:             set E_ij as the edge weight;
24:          else
25:             set the edge weight as infinity;
26:          end if
27:       end for
28:    end for
29:    M ← minimum-cost-bipartite-matching for nodes in G_BM;
30:    for all e_ij in the matching M do
31:       mark u_ij as scheduled; X_i ← the matching F_j;
32:       E_total ← E_ij + E_total;
33:       SL_{X_i} ← max(EST_i, SL_j) + t_{X_i}(i);
34:       Add the data migration delay into SL_{X_i};
35:       for all u_k which is a dependent of u_i do
36:          update the dependence information for u_k;
37:          if EST_k < SL_{X_i} then
38:             EST_k ← SL_{X_i};
39:          end if
40:       end for
41:    end for
42: end while
```

est start time is considered: if it has no dependency constraint at that time, it's inserted into the ready list. A bipartite matching graph is constructed as follows: all nodes from the ready list are on one side, denoted by a set V_1, and all sensors are on the other side, denoted by a set V_2. Each node, u_i, in V_1 has an edge connected with each sensor, F_j, in V_2. If the schedule length of the sensor plus the node's computation time is less than the finish time of the node in ALAP, the edge weight is set to the energy consumption E_{ij}. Otherwise, the edge weight is set to be infinity. After constructing the graph, call the minimum-cost-bipartite-matching function to get a minimum cost bipartite matching. Since the edge weight is set to the energy consumption, the matching produced by the function minimizes the reliability cost in each scheduling step. After a match is found, schedule the tasks on the corresponding sensors, mark the node and update their descendants' information, i.e., dependence constraints and earliest starting time, recursively. Repeat this process until there is no more node to be rescheduled. Because the sensor selection for a node is limited by the finish time in the ALAP, the node can at least be scheduled on the same sensor as ALAP. Thus, as long as there is an ALAP schedule for the graph, this algorithm won't fail to produce a schedule for all the tasks. As the energy is the matching factor in the bipartite matching, the energy consumption minimization is improved over the list scheduling algorithm.

8.3.1.2 Sensor Scheduling with Progressive Relaxation

Sensor_Sch2 progressively improves the reliability based on the schedule obtained by the ALAP scheduling. The idea is to reschedule each node to reduce the system energy consumption as much as possible. It is shown in Algorithm 8.3.2. After the initialization, this algorithm first obtains an ALAP schedule for all the tasks in the graph. Then repeat the following steps until all nodes are marked: among all nodes that are not marked, take the node with earliest starting time and reschedule it to a sensor such that the system energy consumption is minimized. A node can only be rescheduled to a sensor if its finish time is earlier than that in ALAP and the task does not overlap with other tasks remaining in the ALAP schedule. The choice of the node is made this way because the remaining nodes can always be scheduled within the time constraint as long as the ALAP schedule exists for this task graph. After this task is scheduled, mark the node, update the dependence constraint information and the earliest start time for all its descendants recursively, update the system energy consumption and the schedule length, then continue this process until all nodes are marked.

8.3.2 Adaptive Energy-Aware Online Algorithm

In this subsection, we will propose our algorithms to solve the energy-saving problem of heterogeneous virtual sensor networks. The basic idea is to use a time interval (window) and obtain the best mode assignment of

Algorithm 8.3.2 Sensor_Sch2 Algorithm

Require: a DAG $G = \langle V, E_d \rangle$, a set of sensors, a timing constraint L
Ensure: A task scheduling with failure-rate minimization

```
1: for all u_i ∈ V; do
2:     X_i ← -1;
3:     EST_i ← starting time of node i in ASAP;
4:     LST_i ← starting time of node i in ALAP;
5:     FT_i ← Finish time of node i in ALAP;
6: end for
7: for all F_j ∈ F do
8:     SL_j ← 0;
9: end for
10: E_total ← 0;
11: while ∃ nodes not marked do
12:     Min ← ∞ ;
13:     take the node u_i, with minimum LST_i and not marked;
14:     for all F_j ∈ F do
15:         compute E_ij;
16:         if SL_j + t_j(i)+ data migration delay ≤ FT_i then
17:             if E_ij < Min then
18:                 Min ← E_ij ; X_i ← F_j;
19:             end if
20:             if E_ij = Min then
21:                 X_i ← the sensor with earlier finish time;
22:             end if
23:         end if
24:     end for
25:     mark u_ij as scheduled;
26:     E_total ← Min + E_total;
27:     SL_{X_i} ← max(EST_i, SL_j) + t_{X_i}(i);
28:     Add the data migration delay into SL_{X_i};
29:     for all u_k which is a dependent of u_i do
30:         update the dependence information for u_k;
31:         if EST_k < SL_{X_i} then
32:             EST_k ← SL_{X_i};
33:         end if
34:     end for
35: end while
```

each node in the time interval (window). Then adjust the mode of each node accordingly online. We proposed an Adaptive Online Energy Saving (AOES) algorithm to reduce energy consumption while satisfying performance requirements for virtual sensor networks. In this algorithm, we use *MAP_Opt* sub-algorithm to give the best mode assignment for each sensor node.

8.3.2.1 AOES Algorithm

Algorithm 8.3.3 Adaptive Online Energy-Saving Algorithm (*AOES*)

Require: A sensor network, R different mode types, and the timing constraint L.

Ensure: a mode assignment to minimize energy E while satisfying L for the sensor network

1: Collect data and predict the PDF of execution time of each node in a time interval (window).
2: Obtain the best mode assignment A by using *MAP_Opt* for each node in the time interval (window).
3: Output results: A and E_{min}.
4: Use online architectural adaptation to reduce energy consumption while satisfying timing constraints with guaranteed probability.
5: Repeat the above steps.

The AOES algorithm is shown in Algorithm 8.3.3. In AOES algorithm, we use the adaptive model to solve energy-saving problems for heterogeneous sensor networks. The adaptive approach includes three steps: First, collect updated information of each node and predict the PDF of execution time of each node in a time interval (window). Second, in each time interval (window), obtain the best mode assignment for each node during the time interval (window) to minimize the energy consumption while satisfying timing constraint with guaranteed probability. Third, use an on-line architecture adaptation control policy. Since our design is for a non-stationary environment, the control policy varies with the environment but is stationary within a time interval (window).

8.3.2.2 MAP_CP Algorithm

In this subsection, we first design a heuristic algorithm for sensor network. We call this algorithm as *MAP_CP*.

The *MAP_CP* algorithm is shown in Algorithm 8.3.4. A *critical path* (CP) of a DAG is a path from source to its destination. To be a legal assignment for a PDFG, the execution time for any critical path should be less than or equal to the given timing constraint. In algorithm *MAP_CP*, we only consider the hard execution time of each node, that is, the case when the probability of the random variable T equals 1. This is a heuristic solution for hard real-time

Algorithm 8.3.4 Heuristic Algorithm for the MAP Problem When the PDFG Is DAG (*MAP_CP*)

Require: R different mode types, a DAG, and the timing constraint L
Ensure: A mode assignment to minimize energy while satisfying L

1: Assign the lowest energy type to each node and mark the type as assigned.
2: Find a CP that has the maximum execution time among all possible paths based on the current assigned types for the DAG.
3: For every node u_i in CP,
4:     For every unmarked type p,
5:         change its type to p,
6:         $r = cost\_increase/time\_reduce$
7:         select the minimum r.
8:         if $(T > L)$
9:             continue
10:         else
11:             exit   /* the best assignment */

systems. We find the CP with minimized energy consumption first, then adjust the energy of the nodes in CP until the total execution time is less than or equal to L.

8.3.2.3 MAP_Opt Algorithm

For sensor networks, we propose our algorithm, *MAP_Opt*, which is shown as follows.

Require: R different modes, a DAG, and the timing constraint L
Ensure: An optimal mode assignment

1. Topological sort all the nodes, and get a sequence A.

2. Count the number of multi-parent nodes t_{mp} and the number of multi-child nodes t_{mc}. If $t_{mp} < t_{mc}$, use bottom up approach; otherwise, use top down approach.

3. For bottom up approach, use the following algorithm. For top down approach, just reverse the sequence. $|V| \leftarrow N$, where $|V|$ is the number of nodes.

4. If the total number of nodes with multi-parent is t, and there are maximum K variations for the execution times of all nodes, then we will give each of these t nodes a fixed assignment.

5. For each of the K^t possible fixed assignments, assume the sequence after topological sorting is $u_1 \rightarrow u_2 \rightarrow \cdots \rightarrow u_N$, in bottom up fashion. Let $D_{1,j} = B_{1,j}$. Assume $D^{'}_{i,j}$ is the table that stored minimum total

energy consumption with computed confidence probabilities under the timing constraint j for the sub-graph rooted on u_i except u_i. Nodes $u_{i_1}, u_{i_2}, \cdots, u_{i_w}$ are all child nodes of node u_i and w is the number of child nodes of node u_i, then

$$D^{'}_{i,j} = \begin{cases} (0,0) & \text{if } w = 0 \\ D_{i_1,j} & \text{if } w = 1 \\ D_{i_1,j} \oplus \cdots \oplus D_{i_w,j} & \text{if } w \geq 1 \end{cases} \tag{8.1}$$

6. Then, for each k in $B_{i,k}$.

$$D_{i,j} = D^{'}_{i,j-k} \oplus B_{i,k} \tag{8.2}$$

7. For each possible fixed assignment, we get a $D_{N,j}$. Merge the (Probability, Energy) pairs in all the possible $D_{N,j}$ together, and sort them in ascending to sequence according to probability.

8. Then remove redundant pairs. Finally, get $D_{N,j}$.

In algorithm *MAP_Opt*, we exhaust all the possible assignments of multi-parent or multi-child nodes. Without loss of generality, assume we use bottom up approach. If the total number of nodes with multi-parent is t, and there are maximum K variations for the execution times of all nodes, then we will give each of these t nodes a fixed assignment. We will exhaust all of the K^t possible fixed assignments. Algorithm *MAP_Opt* gives the optimal solution when the given PDFG is a DAG. In Equation (8.1), $D_{i_1,j} \oplus D_{i_2,j}$ is computed as follows. Let $G^{'}$ be the union of all nodes in the graphs rooted at nodes u_{i_1} and u_{i_2}. Travel all the graphs rooted at nodes u_{i_1} and u_{i_2}. For each node a in $G^{'}$, we add the energy consumption of a and multiply the probability of a to $D^{'}_{i,j}$ only once, because each node can only have one assignment and there is no assignment conflict. The final $D_{N,j}$ we get is the table in which each entry has the minimum energy consumption with a guaranteed confidence probability under the timing constraint j.

In algorithm *MAP_Opt*, there are K^t loops and each loop needs $O(|V|^2 * L * R * K)$ running time. The complexity of *Algorithm MAP_Opt* is $O(K^{t+1} * |V|^2 * L * R)$. Since t_{mp} is the number of nodes with multi-parent, and t_{mc} is the number of nodes with multi-child, then $t = min(t_{mp}, t_{mc})$. $|V|$ is the number of nodes, L is the given timing constraint, R is the maximum number of modes for each node, and K is the maximum number of execution time variation for each node. The experiments show that algorithm *MAP_Opt* runs efficiently.

Three Methods Comparison with 5 Sensors									
Exper.	N.	Med.1	Med.2	Med.3					
		E (μJ)	E (μJ)	E(1.0) (μJ)	E(0.9) (μJ)	% M1 (%)	% M2 (%)	% (1.0) (%)
exp1	32	2712	2301	2143	1819	32.9	20.9	15.1
exp2	41	2754	2315	2172	1841	33.2	20.5	15.2
exp3	48	2820	2398	2215	1882	33.3	21.5	15.0
exp4	56	2947	2511	2333	1979	32.8	21.2	15.2
exp5	58	3028	2573	2367	2036	32.8	20.9	14.0
exp6	62	3124	2669	2482	2103	32.7	21.2	15.3
exp7	77	3211	2718	2527	2143	33.3	21.2	15.2
exp8	85	3526	2987	2776	2367	32.9	20.8	14.7
exp9	92	3677	3124	2901	2476	32.7	20.7	14.7
exp10	98	3825	3210	2978	2548	33.4	20.6	14.4
Average Reduction (%)						33.0	20.9	14.9

TABLE 8.4: The comparison of total energy consumption with three methods on various task graphs when the time constraint is 2000.

8.4 Experiments

In this section, we conduct experiments with the algorithms on a set of different task graphs. We build a simulation framework to evaluate the effectiveness of our approach. K different FU types, $F_1, \cdots, F_K$, are used in the system, in which a FU with type F_1 is the quickest with the highest energy consumption and a FU with type F_K is the slowest with the lowest energy consumption. Each task node has different energy consumption under different sensors.

We conducted experiments on three methods: Method 1: list scheduling with *MAP_CP*; Method 2: our algorithm FU_Sch2 with *MAP_CP*. Method 3: our algorithm Sensor_Sch2 with *MAP_Opt* (two cases: soft real-time 0.9 and hard real-time). In the list scheduling, the priority of a node is set as the longest path from this node to a leaf node [134]. The experiments are performed on a Dell PC with a P4 2.1 G processor and 512 MB memory running Red Hat Linux 9.0.

The experimental results for the three methods are shown in Table 8.4 to Table 8.6 when the number of sensors is 5 and 10, respectively. Column "Exper." stands for the experimental task graph we used in the experiments. Column "N." represents the number of nodes of each task graph. Columns "Med.1" to "Med.3" represents the four methods we used in the experiments. Column "E" represents the minimum total system energy consumption obtained from four different algorithms. Columns "% M1" and "% M2" under

Four Methods Comparison with 8 Sensors								
Exper.	N.	Med.1	Med.2	Med.3				
		E (μJ)	E (μJ)	E(1.0) (μJ)	E(0.9) (μJ)	% M1 (%)	% M2 (%)	% (1.0) (%)
exp1	32	3793	3163	2945	2470	34.9	21.9	16.1
exp2	41	3864	3217	2976	2507	35.1	22.1	15.8
exp3	48	3947	3298	3073	2572	34.8	22.0	16.3
exp4	56	4117	3410	3178	2671	35.1	21.7	16.0
exp5	58	4254	3557	3286	2769	34.9	22.2	15.7
exp6	62	4379	3652	3398	2847	35.0	22.0	16.2
exp7	77	4502	3729	3471	2921	35.1	21.7	15.8
exp8	85	4918	4088	3790	3187	35.2	22.0	15.9
exp9	92	5140	4279	3998	3352	34.8	21.7	16.2
exp10	98	5358	4462	4134	3477	35.1	22.1	15.9
Average Reduction (%)						35.0	21.9	16.0

TABLE 8.5: The comparison of total energy consumption with four methods on various task graphs when the time constraint is 3000.

"Med.3" represent the percentage of reduction in total energy consumption, compared to the Method 1 and Method 2, respectively. The average reduction is shown in the last row of the table.

The results show that our algorithms can significantly improve the performance of applications with one sensor networks. Our algorithm (Method 3) improves the energy consumption reduction over the traditional list scheduling algorithm and *MAP_CP*. Among them, Method 3 (0.9) gives the best performance. We can see that with more sensors selections, the reduction ratio for the total energy consumption has increased. For example, with five sensors, compared with Method 1, Method 3 (0.9) shows an average 33.0% reduction in total energy consumption. While using 10 sensors, the reduction rate changed to be 38.1% for total energy consumption.

It is worthwhile to point out that we obtain this improvement ratio without sacrificing performance. Sensor_Sch1 improves the energy consumption reduction significantly when the time constraint is large. If the time constraint is small, it still improves the energy consumption reduction while meeting the constraint. Sensor_Sch1 can always improve energy consumption reduction as long as there exists a schedule by the ALAP scheduling. The improvement is not as significant as Sensor_Sch2 when the time constraint is large. This is because the algorithm always tries to use all available sensors in each step, while Sensor_Sch2 schedules one node in each step and avoids sensors with high energy consumption if possible.

Four Methods Comparison with 10 Sensors								
Exper.	N.	Med.1	Med.2	Med.3				
		E (μJ)	E (μJ)	E(1.0) (μJ)	E(0.9) (μJ)	% M1 (%)	% M2 (%)	% (1.0) (%)
exp1	32	5154	4218	3917	3201	37.9	24.1	18.3
exp2	41	5235	4276	3967	3254	37.8	23.9	18.0
exp3	48	5372	4395	4089	3342	37.8	24.0	18.3
exp4	56	5610	4527	4175	3436	38.8	24.1	17.7
exp5	58	5742	4662	4328	3552	38.1	23.8	17.9
exp6	62	5926	4845	4487	3671	38.1	24.2	18.2
exp7	77	6121	4965	4645	3791	38.1	23.6	18.4
exp8	85	6678	5433	5011	4129	38.2	24.0	17.6
exp9	92	6992	5701	5297	4339	37.9	23.9	18.1
exp10	98	7278	5955	5503	4517	37.9	24.1	17.9
Average Reduction (%)						38.1	24.0	18.0

TABLE 8.6: The comparison of total energy consumption with four methods on various task graphs when the time constraint is 4000.

8.5 Conclusion

In this chapter, we present some optimization methods in networked sensing and monitoring system designs. We proposed two highly efficient algorithms to solve task-sensor scheduling problems for sensor network applications. By using multiple sensors scheduling, our algorithms improve both the performance and energy minimization of sensor networks. These algorithms not only are optimal, but also provide more choices of smaller total energy consumption with guaranteed confidence probabilities satisfying timing constraints. In many situations, algorithm MAP_CP cannot find a solution, while ours can find satisfied results.

8.6 Glossary

Wireless Sensor Network: A wireless sensor network (WSN) consists of spatially distributed autonomous sensors to cooperatively monitor physical or environmental conditions, such as temperature, sound, vibration, pressure, motion or pollutants.

Chapter 9

Battery-Aware Scheduling for Wireless Sensor Network

9.1 Introduction

Wireless sensor network (WSN) has recently received tremendous attention. The interest is growing due to the benefits WSN brings and the large number of unexplored applications. However, WSN faces challenges which limit their usability. One of the most notable is the energy limit. The wireless sensor node can only be equipped with a limited power source (< 0.5 Ah, 1.2V) [8]. In some application scenarios, replenishment of power resources might be impossible. In these cases, sensor node lifetime strongly depends on battery lifetime. In most of WSN, sensor nodes play as not only data originators but data routers. The loss of some sensor nodes in WSN may cause significant overhead of network topological reorganization and rerouting. Meanwhile, many emerging applications require processing which has considerable computation demands and needs to be done in the sensor network. For instance, in target tracking system [214], sensors collaboratively measure and estimate the location of moving targets or classify targets. Another typical application is the video sensor network [71]. Since the size of raw video data is too large to transfer in the network, sensors are required to process some computationally intensive operations.

Parallel processing in WSN can be a solution to intensive computation requirement. Some problems need to be solved when applying parallel processing: 1) how to assign tasks to the processing units in the sensors; 2) in what order the processing units should execute the tasks assigned to it; and 3) how to schedule communication among the network. Task scheduling can solve these three problems. Task scheduling has been studied in high performance computing [65, 83]. However, a useful scheduling algorithm strongly depends on the accuracy of the model it is based on. Applying task scheduling in WSN, we need to develop a model which is different from most of the high computing architecture models. Besides, task scheduling in WSN should be subject to some limitations of WSN, for instance, power consumption, lifetime requirement and so on.

Due to the fact that most of the wireless sensors are equipped with batteries, battery behavior modeling is significant when describing the power consumption precisely in WSN. Recent research shows that the discharging loss, one kind of energy loss when keeping the battery on without any "sleep" periods, can be as high as 30% of the total energy [126]. Therefore a battery-aware task scheduling algorithm should be more suitable for parallel processing in WSN.

The two major contributions of this chapter are:

- We present a complete model for task scheduling in WSN, which includes application model, network model as well as battery model.
- We propose a two-phase list-scheduling algorithm for scheduling tasks. It can generate a schedule with close-to optimal total execution time while subject to the battery lifetime constraint.

In this chapter, we propose a battery-aware task scheduling algorithm for WSNs. Our proposed solution, two-phase list-scheduling algorithm, aims to schedule the task to wireless sensors with minimum total execution time, and subject to battery lifetime constraint. In Section 9.2, we discuss works related to this topic. In Section 9.3, models for task scheduling in WSN are presented. We propose our algorithm in Section 9.4, followed by experimental result in Section 9.5. Finally, we give the conclusion in Section 9.6.

9.2 Related Work

Task scheduling in WSN has been studied in the literature recently. In [20], the authors present a task mapping approach for mobile ad hoc networks. And a data fusion task mapping approach is presented in [108]. These two approaches both assume an existing network communication mechanism. They do not solve the problem of scheduling communication in network. In [80], an online scheduling mechanism is proposed to allocate resources to periodic task. Authors in [206] propose an energy-aware task scheduling algorithm, EcoMapS. Similar to our proposed algorithm, EcoMapS is based on list-scheduling. But the WSN concern in EcoMapS is homogenous sensor network. This limits its applicability. Study in [188] focuses on heterogeneous mobile ad hoc grid environments. However, the communication model in this study is not suitable for WSN. And none of the works above are concerned about the battery behavior in WSN.

The experiment conducted by Rakhmatov and Vrudhula [173] shows that the energy dissipated in the device is not equivalent to the energy consumed from a battery. When discharging, the energy consumed in the battery is more than needed. And in idle time, the over-consumed energy is recovered. Several analytical models on battery discharging behavior have been developed recently [148, 173, 126]. In [148], Panigrahi provided a model based on a negative exponential function, and the discharging and recovery are represented as a transient stochastic process. Rakhmatov and Vrudhula [173] proposed an analytical battery model based on a one-dimensional model of diffusion in a finite region. However, these two models are not suitable for task-scheduling in WSN due to their high computational complexity. Ma presents an online computable battery model in [126]. The relatively low computational complexity makes it suitable for task-scheduling. The battery model used in this chapter is based on Ma's model.

9.3 Background and Model

9.3.1 Application Model

In this chapter, we use the *Directed Acyclic Graphs* (DAG) to represent applications. A DAG $T = (V, E)$ consists of a set of vertices V, each of which represents a task in the application, and a set of edges E, showing the dependencies among the tasks. The edge set E contains edges e_{ij} for each task $v_i \in V$ that task $v_j \in V$ depends on. The weight of a task represents the task type of this task. And the weight of an edge e_{ij} means the size of data which is required by v_j and produced by v_i.

Given an edge e_{ij}, v_i is the immediate predecessor of v_j, and v_j is called the immediate successor of v_i. A task only starts after all its immediate predecessors finish. Tasks with no immediate predecessor are entry-tasks, and tasks without immediate successors are exit-tasks.

9.3.2 Network Model

Usually, a large number of sensors are deployed throughout the sensor field as shown in Figure 9.1. Each of the scatted sensors has the capability to collect data and route data to the sink. The sink is the network node acting as a cluster head, communicating with the task manager node via Internet or Satellite [8]. All the identical sensors collaborate as a cluster. The sensor network consists of different types of sensors with different computation power as well as data collecting and transmitting capacity. All these sensors work as a set of clusters. Therefore, from the whole sensor network standpoint, the network we focus on is a heterogeneous sensor network. The following assumptions are made for the WSN:

- Clusters are formed by identical sensors. We assume that all the sensors in the same cluster have the same computation and communication capacities, the same battery behaviors. And sensors are multi-hop connecting to the sink [8].

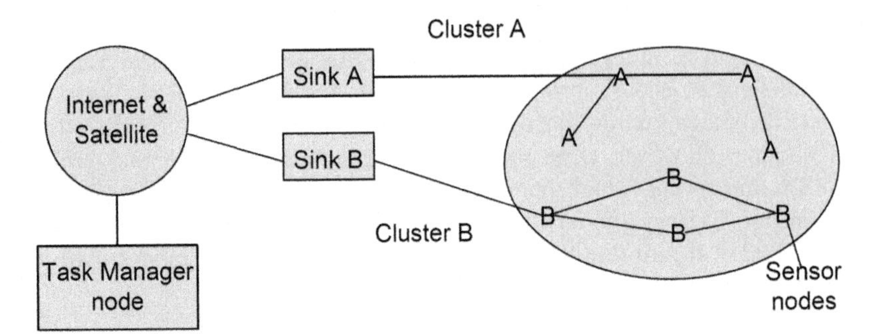

FIGURE 9.1: An example of wireless sensor network.

- Tasks are assigned by the tasks manager node through the Internet to the sinks of clusters. We assume the bandwidths of Internet connection among task manager node and sinks are the same and the task manager node is able to communicate data with different nodes simultaneously. The bandwidth of network inside the cluster is different from cluster to cluster.
- Computation and communication can occur simultaneously on sensors.
- When sensors send the result back to sink, they keep power on until all the sensors in the cluster have their result back to the sink, which means in a task, all sensors in the cluster will turn off at the same time. We make this assumption based on the following reason: all the sensors in the cluster are identical; the execution time of measurement and computing procedure on each sensor should be the same, which means all the results can be ready at the same time. What's more, because of the multi-hop network in the cluster, sensors should keep power on in order to route data to other sensors to the sink.

Here is an example of how the WSN assigns, executes task and collects the result. First of all, the task manager node assigns the task to a cluster. Meanwhile, the required result data of all the predecessors of the current task are sent from manager node to the sink of that cluster. And this communication time T_{init_comm} is proportional to the total size of predecessor result data M_{pred}:

$$T_{init\_comm} = \frac{M_{pred}}{BW_{internet}} \tag{9.1}$$

$BW_{internet}$ is the Internet bandwidth. Only when the predecessor task was executed in the same cluster as the current task, this communication is not required because of the already existing data. Once the sink receives all the predecessor result data, it broadcasts the measure command to the sensors in its cluster. Then the sensors turn on their power and start the measurement and the computing procedure. The total time of these two procedures T_{sensor} depends on the processing speed of executing that kind of task on the assigned sensor SP_{sensor} and the size of the task data M_{td}:

$$T_{sensor} = SP_{sensor} \times M_{td} \tag{9.2}$$

After computing the result data, sensors send their result back to the sink. The total time for collecting all the results in a cluster T_{result} ideally should be proportional to the product of the size of result data M_{result} and the network bandwidth within the cluster $BW_{cluster}$:

$$T_{result} = \frac{M_{result}}{BW_{cluster}} \tag{9.3}$$

When collecting the results is done, all the sensors turn their power off and wait for the next assignment. The sink sends the result data back to the task manager node. This communication time $T_{result_comm}$is similar to T_{init_comm}:

$$T_{result\_comm} = \frac{M_{result}}{BW_{internet}} \tag{9.4}$$

Because the sensors are identical in a cluster—they turn on at the same time, compute with the same speed, and turn off together—we can consider that the behaviors of the cluster represent the behaviors of the sensors in it. An *execution speed to compute* (EPC) matrix is known as prior. An entry e_{ij} is the execution speed of task t_i on sensor cluster c_j. So the execution time of task t_i on sensor cluster c_j is $e_{ij} \times M_{task}$, where $M_{task}(t_i)$ is the data size of task t_i.

9.3.3 Battery Behavior and Energy Constraint

Nickel-cadmium and lithium-ion batteries are the most commonly used batteries in wireless sensor networks. These kinds of batteries consist of an anode and a cathode, separated by an electrolyte. When connected to a load, a reduction-oxidation reaction transfers electrons from the anode to the cathode. Active species are consumed at the electrode surface and replenished by diffusion from the bulk of the electrolyte. However, this diffusion process cannot keep up with the consumption. And a concentration gradient builds up across the electrolyte. When this concentration falls, the battery voltage drops. And when the voltage is below a certain cutoff threshold, the electrochemical reaction cannot be sustained at the electrode surface anymore, so the battery stops working. But in fact, the active species which has not yet reached the electrode are not used. This unused charge is called discharging loss. Discharging loss is not physically lost but simply unavailable. If the battery current is reduced to a low value or even zero before the battery stops working, the concentration gradient flattens out after a sufficiently long time, and the remaining active species reach the electrode again. Then the discharging loss is available for extraction. This procedure is called the battery recovery [126]. Experiments show that this discharging loss might take up to 30% of the total battery capacity [126].

Precisely modeling battery behavior is essential for optimizing system performance. The battery behavior model used in this chapter is based on Ma's approach [126]. Consider the scenario where a battery is turned on for δ_i time, and turned off for τ_i time ($i = 1, 2, \ldots$). This on-off period is repeated until the battery dies. We assume that the discharging current of the battery in epoch δ_i is I_i, and the beginning time of this epoch is t_i. The energy dissipated by the battery in epoch δ_i is:

$$\Delta\alpha = I_i \times \delta_i + 2I_i \times \sum_{m=1}^{\infty}[\frac{e^{-\beta^2 m^2(T-(t_i+\delta_i))} - e^{-\beta^2 m^2(T-t_i)}}{\beta^2 m^2}] \quad (9.5)$$

The model is interpreted as follows. The first term in the right-hand side of (9.5) is simply the energy consumption during the epoch δ_i. And the second term is the discharging loss during the δ_i epoch. T is the entire lifetime of the battery when the battery is on until it dies (greedy mode). β is a positive constant, which is determined in experiment and may vary from battery to battery.

An idle period τ_i follows the epoch δ_i. In this idle period, the battery "sleeps" with very low current on it. In the WSN scenario, the battery is turned off when the cluster has finished the current task and is waiting for the next task. The residual discharging loss when it is t time after epoch δ_i can be computed as:

$$\zeta_i(t) = 2I_i \times \sum_{m=1}^{\infty}[\frac{e^{-\beta^2 m^2(T+t-(t_i+\delta_i))} - e^{-\beta^2 m^2(T+t-t_i)}}{\beta^2 m^2}] \quad (9.6)$$

$\zeta_i(0)$ equals to the discharging loss of δ_i. Note that this residual discharging loss is just a potential energy in the sense that it only makes sense when the battery is alive. Once the battery dies, this residual energy will not be recovered. When the battery is alive during the τ_i period, the energy recovered at the end of the τ_i period is:

$$\begin{aligned}\Delta\alpha_r(\tau_i) = \quad & 2I_i \times (\textstyle\sum_{m=1}^{\infty}[\frac{e^{-\beta^2 m^2(T-(t_i+\delta_i))} - e^{-\beta^2 m^2(T-t_i)}}{\beta^2 m^2}] \\ & - \textstyle\sum_{m=1}^{\infty}[\frac{e^{-\beta^2 m^2(T+\tau_i-(t_i+\delta_i))} - e^{-\beta^2 m^2(T+\tau_i-t_i)}}{\beta^2 m^2}]) \end{aligned} \quad (9.7)$$

In WSN, we assume that sensors are equipped with batteries which cannot be replenished. Because sensors in the same cluster are identical, they have the same size of battery capacity; for instance, the battery capacity of each sensor in cluster i is E_i(mAmin). Without loss of generality, we assume that the discharge current CUR_{ij}(A) of each sensor in the cluster i running task j is different from cluster to cluster, task to task. We also assume the discharge current of a sensor in cluster i transmitting data is CUR_T_i. Given a certain task mapping and schedule, we can calculate the energy consumption of the sensors in the clusters with equation (9.5) and (9.7). If all the clusters can finish all the tasks assigned to them before the batteries die in a given schedule, obviously this schedule does not violate the lifetime. However, in some cases, some clusters may not be able to finish tasks before the batteries die. Then the battery lifetime of these clusters can be calculated. If any of these lifetimes is shorter than a predetermined lifetime constraint $C_{lifetime}$, this schedule violates the lifetime constraint.

Figure 9.2 shows an example of how the battery lifetime constraint impacts the WSN schedule. Figure 9.2(a) is the application DAG implemented in the example. There are two clusters in the WSN. Figure 9.2(b) and Figure 9.2(c) are two different schedules. Obviously, the first schedule has the shorter finish time than the second one. But unfortunately, the batteries of the cluster A die at time 6 (the horizontal line across the rectangle 2); the application is incomplete. For the second schedule, even though the finish time is a little longer, both clusters satisfy the battery lifetime constraint.

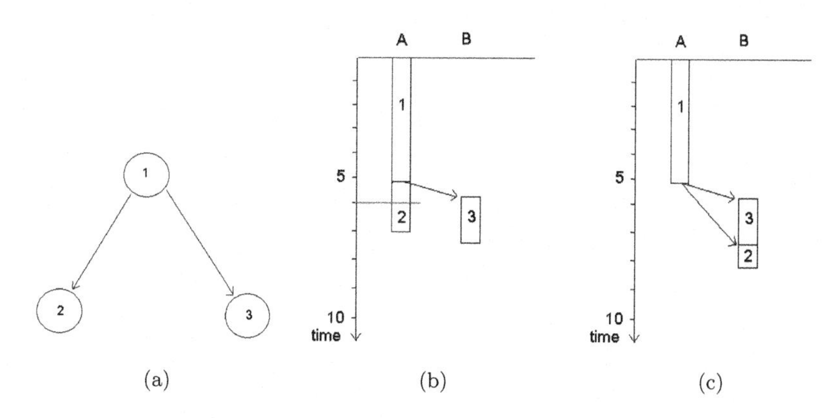

FIGURE 9.2: An example of WSN schedules under battery lifetime constraint. (a) A simple DAG. (b) and (c) Two different schedules for the DAG.

9.4 Two-Phase List-Scheduling Algorithms

List-scheduling is widely used as a static scheduling approach for DAG-based applications with low complexity and good result. List-scheduling first forms an order of jobs by assigning priority to each job. Then jobs are selected in the order of their priorities. In each step, the selected job is assigned to the processor or machine which minimizes the predefined objective function. In this chapter, we set the objective function to the time when WSN finishes all the assigned tasks. Additionally, in order to satisfy the lifetime constraint, the battery lifetimes of all the clusters are calculated. Once the constraint is violated, a reassigning is conducted to obtain an optimal schedule subject to the lifetime constraint.

9.4.1 Phase I: DAG List Scheduling

The list scheduling used in Phase I is similar to CPNT [83]. The definitions are provided as follow. *Cluster available time* (CAT) is the time when the cluster finishes all the tasks which are previously assigned to this cluster. *Task available time* (TET) is the time when all the predecessors of this task are finished and have their result data sent back. These two definitions are based on the scheduling decisions made in the previous steps of the algorithm. Some definitions used in listing the task are provided as follows. The *earliest start time* (EST) of task v_i is the ideal earliest start time of v_i based on the DAG information. The entry-tasks have EST $= 0$. And EST of each vertex is calculated by traversing the DAG from the top to the bottom. Similarly, the *latest start time* (LST) of task v_i is the latest start time so that the execution of v_i's successors will not be delayed. The LST of exit-tasks equal their EST. The LST of each vertex is calculated by traversing the DAG from the bottom to the top. EST and LST of a task v_i are calculated as follows:

$$EST(v_i) = \max_{v_m \in pred(v_i)} \{EST(v_m) + AT(v_m)\} \tag{9.8}$$

$$LST(v_i) = \min_{v_m \in succv_i} \{LST(v_m)\} - AT(v_i) \tag{9.9}$$

Because the WSN concerned in this chapter is heterogeneous, the execution time of a task on different clusters is not the same. $AT(v_i)$ is the average execution time of task v_i. The critical node (CN) is a set of vertices in the DAG of which EST and LST are equal. Figure 9.3 shows a function forming a task list based on the priorities.

Once the list of task is formed, we can assign tasks to clusters in the order of this list. The task on the top of the list is assigned to the cluster which can finish it at the earliest time. Then this task is removed from the list. The procedure repeats until the list is empty. An optimal schedule is obtained after this assigning procedure which is shown in Figure 9.4.

9.4.2 Phase II: Constraint-Aware Reassigning

To satisfy the lifetime constraint, we need to conduct a reassigning if the schedule obtained in the previous phase violated the lifetime constraint. First of all, we examine the battery lifetimes of all clusters. Clusters violating the lifetime constraint, which are called urgent clusters, will be pushed into a list. We try to reassign the tasks in the urgent clusters to other cluster, until urgent

Require: A DAG, Average execution time AT of every tasks in the DAG
Ensure: A list of tasks P based on priorities
1: The EST of every task is calculated.
2: The LST of every task is calculated.
3: Empty list P and stack S, and pull all tasks in the list of task U.
4: Push the CN task into stack S in the decreasing order of their LST.
5: **while** the stack S is not empty **do**
6:   **if** $top(S)$ has un-stacked immediate predecessors **then**
7:     $S \leftarrow$the immediate predecessor with least LST
8:   **else**
9:     $P \leftarrow top(S)$
10:     pop $top(S)$
11:   **end if**
12: **end while**

FIGURE 9.3: Forming a task list based on the priorities.

Require: A priority-based list of tasks P, m different sensor clusters, EPC matrix
Ensure: A schedule generated by list-scheduling
1: **while** the list P is not empty **do**
2:   $T = top(P)$
3:   find the cluster C_{min} giving the earliest finish time of T
4:   assign task T to cluster C_{min}
5:   remove T from P
6:   update CET of cluster C_{min} and TET of successors of T
7: **end while**

FIGURE 9.4: The assigning procedure.

Require: A schedule S, battery lifetime constraint CS
Ensure: A schedule satisfies lifetime constraint CS
1: A list of urgent clusters U is generated.
2: **while** the list U is not empty **do**
3: $C_u = top(P)$
4: a list of task L is generated based on top-to-bottom selecting or bottom-to-top selecting.
5: **while** C_u violates the constraint CS **do**
6: $T_{target} = top(L)$
7: reassign(T_{target})
8: remove T_{target} from L
9: **end while**
10: **end while**

FIGURE 9.5: The adjusting procedure.

cluster list is empty. Some definitions are used as follows. Given a schedule, the latest predecessor-finish time of a task v_i LPFT(v_i) is the time when all its predecessors are finished and have all the result data back to the task manager node. The earliest successor-start time of a task v_i ESST(v_i) is the earliest time when any of its successors has already had the v_i result data sent from task-manager and is scheduled to start the execution in a given schedule. The execution zone of v_i is the time between LPFT(v_i) and ESST(v_i). Obviously, execution zone of v_i is the time when v_i is ready to be executed. And even if v_i can't finish the execution as scheduled previously, the execution of its successors won't be delayed as long as v_i finishes in its execution zone. In reassigning phase, a target task is the task to be rescheduled. A target cluster is the cluster to which the target task is rescheduled. Several criteria are used to determine target cluster for a given target task:

1. Target cluster should not be the urgent cluster.

2. Target cluster should be idle in the execution zone of the target task.

3. The clusters with predecessors and/or successors of the target task are preferred when choosing the target cluster.

The idea behind criterion 2 above is that when reassigning the target task to a cluster which is idle in the execution zone of this task, the successors of target task and the following tasks in the task list of target cluster won't be delayed. So the total finish time of this cluster is the same as the original one. And when choosing the target cluster in criterion 3 above, the total finishing time may be shorter, due to less data to communicate. A function of reassigning a given target task to target cluster selected by combination of these three criteria above is shown in Figure 9.6.

When selecting target task from the task-list of an urgent cluster, we use two different schemes. One is top-to-bottom selecting. In this scheme, a list of tasks assigned in urgent cluster is formed in the order of decreasing energy consumption. The top task of the list is selected to reassign. If the lifetime constraint is still violated after assigning the selected task, then the tasks are chosen in the order of that list until the lifetime constraint is satisfied. Similarly, the other one is bottom-to-top selecting. However, in this scheme, the list is formed in the order of increasing energy consumption. Tasks are chosen in the order of that list until constraint is met. We will compare the results of these two different schemes later in this chapter. A function of the complete Phase II algorithm is shown in Figure 9.5.

9.5 Experimental Results

9.5.1 Experiment Setup

We evaluate the performance of the two-phase list-scheduling algorithms through simulations. Each simulation run (10 total) has 64 unique applications, and each application is composed of up to 16 tasks. For a task, the maximum fan-in and fan-out are both 3. There are 32 clusters of sensors in the WSN. We set parameters in the model randomly according the maximum and minimum values shown in Table 9.1. δ of all batteries are set to 0.1. Lifetime constraint for all clusters is set to 450. And Internet bandwidth is set to 10.

parameter	Minimum	Maximum
$EPC_{i,j}$	10	40
$CUR_{i,j}$	20	100
data size	20	148
cluster bandwidth	2	10
E_i(mAmin)	1.0×10^5	8.0×10^5
CUR_T_i	20	400

TABLE 9.1: Range of model parameters.

9.5.2 Result

In Section 9.4, we let the algorithm keep reassigning tasks in urgent cluster until lifetime constraint is met. Since we set the lifetime constraint shorter than the finish time of most clusters, the reassigning cannot guarantee that all clusters can finish all their tasks. So we further set another condition of

```
Require: A schedule S, battery lifetime constraint CS, a target task
Ensure: A reassigning a target task
 1: if clusters which satisfy all three criteria exist then
 2:    a list of cluster CL1 satisfying all three criteria is generated
 3:    while (the target task has not been reassigned yet) and (CL1
       is not empty) do
 4:       calculate the lifetime of cluster top(CL1) assuming target
          task is reassigned to it
 5:       if lifetime constraint is not violate then
 6:          reassign target task to top(CL1)
 7:       else
 8:          remove top(CL1) from CL1
 9:       end if
10:    end while
11: else if clusters which satisfy criteria 1 and any one of 2, 3 exist
    then
12:    a list of cluster CL2 satisfying the condition above is generated
13:    while (the target task has not been reassigned yet) and (CL2
       is not empty) do
14:       calculate the lifetime of cluster top(CL2) assuming target
          task is reassigned to it
15:       if lifetime constraint is not violate then
16:          reassign target task to top(CL2)
17:       else
18:          remove top(CL2) from CL2
19:       end if
20:    end while
21: else if clusters which satisfy only criteria 1 exist then
22:    a list of cluster CL3 satisfying the condition above is generated
23:    while (the target task has not been reassigned yet) and (CL3
       is not empty) do
24:       calculate the lifetime of cluster top(CL3) assuming target
          task is reassigned to it
25:       if lifetime constraint is not violate then
26:          reassign target task to top(CL3)
27:       else
28:          remove top(CL3) from CL3
29:       end if
30:    end while
31: else
32:    exit(fail to reassign target task)
33: end if
```

FIGURE 9.6: A function of reassigning target task to target cluster.

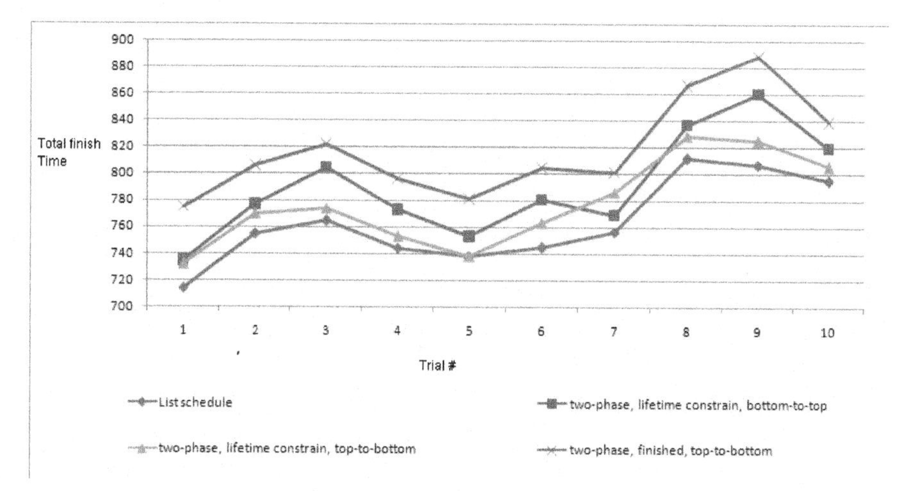

FIGURE 9.7: Total finish time.

reassigning to be when the urgent cluster cannot finish all its tasks. In our experiment, we compare the results of these two different reassigning conditions. Also, the results from original list-scheduling and our two-phase algorithm with those two tasks selection schemes (see Section 9.4) are compared.

In each run, the list-scheduling algorithm first generates an initial schedule, then based on this initial schedule, new schedules are generated by the second phase of our proposed algorithm with different task selection scheme (top-to-bottom or bottom-to-top) and different reassigning condition (lifetime or finishing). From Figure 9.7, we find that the schedules from original list-scheduling have the shortest total finish time. Note that the total finish time is the longest scheduled finish time among all clusters in a given schedule when ignoring batteries lifetimes. Because the original list-scheduling does not consider energy consumption, it has a better result in total finish time. However, in the sense of finish time, the other three schedules generated after reassigning are still close to the list-scheduling one, with less than 10% longer finish time, even though they consider the energy consumption. In the respect of satisfying lifetime constraint, our proposed algorithm does much better than direct list-scheduling, shown in Figure. 9.8. Reassigning with finish condition only fails to avoid violating lifetime constraint in one run. And those two reassigning with lifetime constraint condition pass in all runs. The reason why reassigning with finish condition fails once and those with lifetime constraint condition don't fail is that when using the lifetime constraint condition, the algorithm tries its best to push tasks away from urgent cluster. In some cases it may cause the non-urgent cluster to fail to complete all the tasks within its battery capacity, even though this non-urgent cluster still satisfies the lifetime constraint. But when using the finish condition, the algorithm will let the clusters finish as much as possible of the task, even though in some cases, some

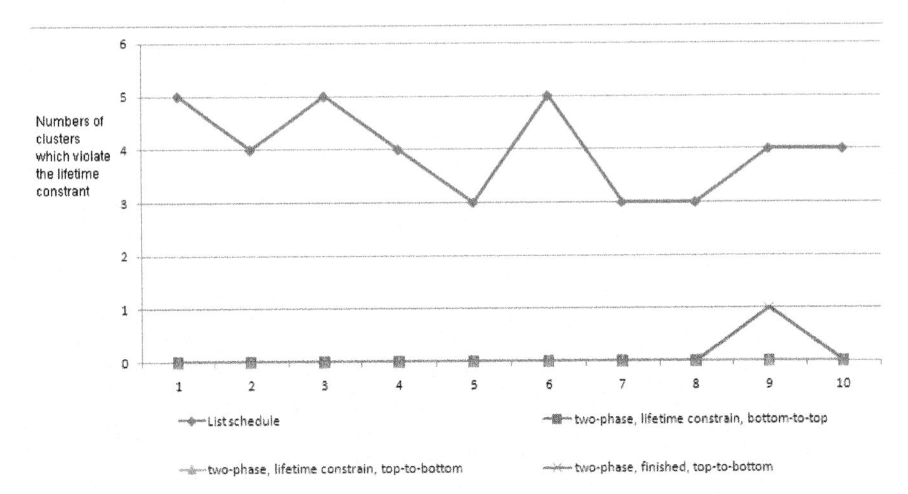

FIGURE 9.8: Numbers of clusters which violate the lifetime constraint.

clusters may still fail in satisfying the lifetime constraint. We think these two kinds of effort are both valuable. In some scenarios, the batteries are scheduled to be changed after a period of time. In this case, the lifetime constraint is more important. In some other scenarios, the batteries are not expected to renew in the future, so the clusters should finish as many tasks as possible. Figure 9.9 shows us that reassigning with lifetime constraint cannot improve the completion of tasks much while reassigning with finish condition can avoid most of the unfinished clusters. Since reassigning with lifetime constraint is pushing the task away from urgent cluster, it may probably generate a large idle time slot in urgent cluster. So the recovered discharging loss in those two schedules is larger than the other two (see Figure 9.10). When comparing the performance of those two task selection schemes, top-to-bottom is better. Schedules generated by top-to-bottom have finish time closest to the ones from original list-scheduling, only about 4% longer on average. The reason is that in a top-to-bottom scheme, the algorithm pushes the task with the largest energy consumption away first. So the task requiring larger energy is more likely reassigned to the best cluster in the non-urgent cluster list in top-to-bottom scheme than in bottom-to-top scheme.

From the system point of view, the cluster with the minimum lifetime has a larger impact on WSN. The shorter lifetime of that cluster, the more frequently WSN needs to reorganize its topology. So the longest the minimum lifetime the schedule has, the better the schedule is. We find that the reassigning with finish constraint and top-to-bottom scheme has the best performance in the sense of minimum lifetime (see Figure 9.11). And all three reassignings can generate longer minimum lifetime than the original list-scheduling.

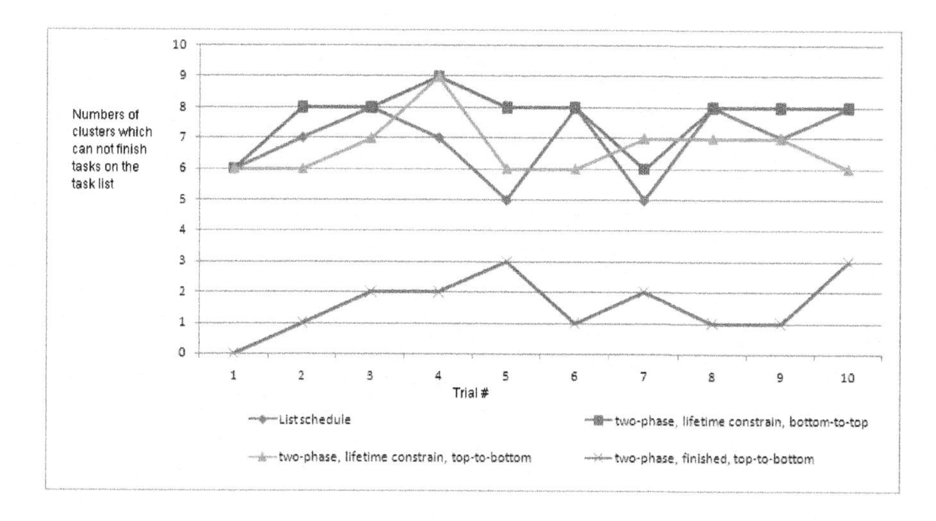

FIGURE 9.9: Numbers of clusters which cannot finish tasks on their task list.

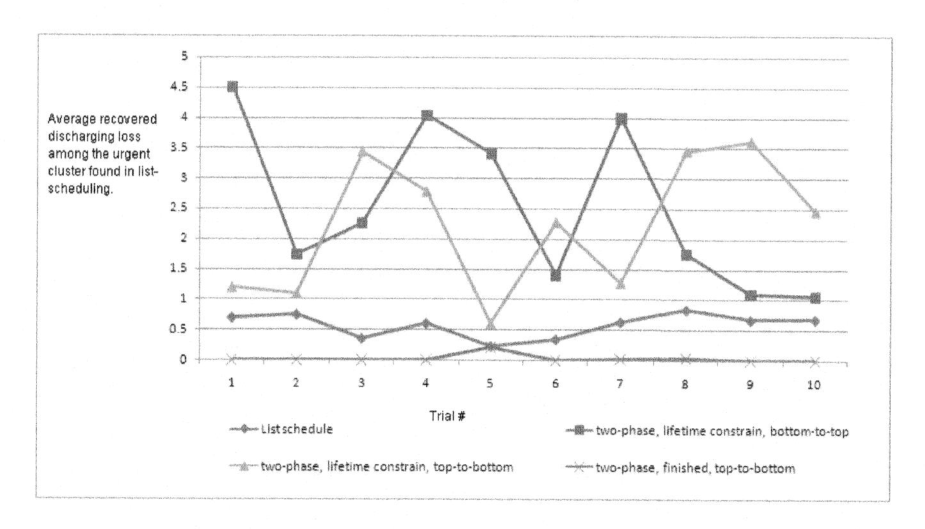

FIGURE 9.10: Average recovered discharging loss among the urgent clusters found in list-scheduling.

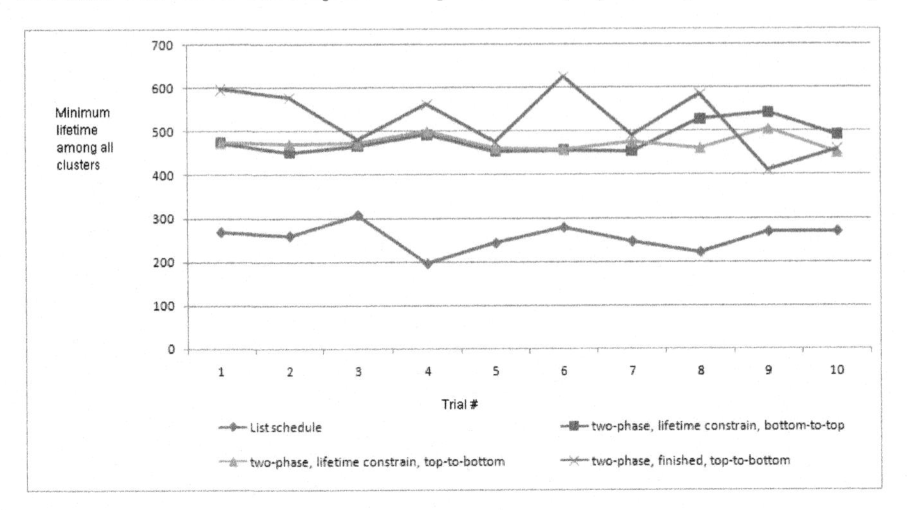

FIGURE 9.11: Minimum lifetime among all clusters.

9.6 Conclusion

In this chapter, we present a complete model for task scheduling in WSN parallel processing, which includes application model, network model as well as battery behavior model. Using this model, we propose a battery-aware task scheduling algorithm, two-phase list-scheduling algorithm. We show that this algorithm can generate close to optimal scheduling while satisfying lifetime constraint. Our further work includes extending our application model and adding sensor failure handling in the algorithm.

9.7 Glossary

Lithium-Ion Batteries: a family of rechargeable battery types in which lithium ions move from the negative electrode to the positive electrode during discharge, and back when charging.

Nickel-Cadmium Batteries: A type of rechargeable battery using nickel oxide hydroxide and metallic cadmium as electrodes.

Chapter 10

Adaptive Resource Allocation in Cloud Systems

Keywords: Cloud computing, adaptive scheduling, feedback, preemptable scheduling

10.1 Introduction

In cloud computing, a cloud is a cluster of distributed computers providing on-demand computational resources or services to the remote users over a network [79]. In an Infrastructure-as-a-Service (IaaS) cloud, resources or services are provided to users in the form of leases. The users can control the resources safely thanks to the free and efficient virtualization solutions, e.g., the Xen hypervisor [195]. One of the advantages of the IaaS clouds is that the computational capacities provided to end-users are flexible and efficient. The virtual machines (VMs) in Amazon's Elastic Compute Cloud are leased to users at the price of 10 cents per hour. Each VM offers an approximate computational power of a 1.2 GHz Opteron processor, with 1.7 GB memory and 160 GB disk space. For example, when a user needs to maintain a database with a certain disk space for a month, he/she can rent a number of VMs from the cloud, and return them after that month. In this case, the user can minimize the costs. And the user can add or remove resources from the cloud to meet peak or fluctuating service demands and pay only the capacity used. To further take advantage of the huge computation capacities the clouds provide, we can execute the intensive computation applications with parallel processing to achieve high performance.

When an application is submitted to the clouds, it is usually partitioned into several tasks. When applying parallel processing in executing these tasks, we need to consider the following questions: 1) how to allocate resources to tasks; 2) in what order the clouds should execute the tasks; and 3) how to schedule overheads when VMs prepare, terminate or switch tasks. Resource allocation and task scheduling can solve these three problems. Resource allocation and task scheduling have been studied in high performance computing

[65] and in embedded systems [163]. However, the autonomic feature within clouds [79] and the VM implementation require different algorithms for resource allocation and task scheduling in the IaaS cloud computing.

The two major contributions of this chapter are:

- We present a resource allocation mechanism in cloud systems, which enables preemptable task scheduling. This mechanism is suitable for the autonomic feature within clouds and the diversity feature of VMs.
- We propose two adaptive algorithms for resource allocation and task scheduling. We consider the resource contention in the task scheduling.

In Section 10.2, we discuss works related to this topic. In Section 10.3, models for resource allocation and task scheduling in IaaS cloud computing system are presented. We propose our algorithms in Section 10.4, followed by experimental results in Section 10.5. Finally, we give the conclusion in Section 10.6.

10.2 Related Work

Computational resource management in cloud computing has been studied in the literature recently. In [68], the authors propose an image caching mechanism to reduce the overhead of loading disk images in virtual machines. The authors of [70] present a dynamic approach to create virtual clusters to deal with the conflict between parallel and serial jobs. In this approach, the job load is adjusted automatically without running time prediction. A system which can automatically scale its share of infrastructure resources is designed in [178]. Another resource sharing system which can trade machines in different domains without infringing autonomy of them is developed in [177]. But all the researches in above works do not consider the preemptable task scheduling. In [194], a suspend/resume mechanism is used to improve utilization of physical resources. The overhead of suspend/resume is modeled and scheduled explicitly. But the VMs model considered in [194] is homogeneous, so the scheduling algorithm is not applicable in heterogeneous VMs models. Study in [188] focuses on scheduling in heterogeneous mobile ad hoc grid environments. However the scheduler algorithms cannot be used in cloud computing.

10.3 Model and Background

10.3.1 Cloud System

In this chapter, we consider an infrastructure-as-a-service (IaaS) cloud system. In this kind of system, several cloud providers participate. These cloud providers deliver basic on-demand storage and compute capacities over the Internet. The provision of these computational resources is in the form of virtual machines (VMs) deployed in a provider's data center. These resources within a provider form a cloud. Virtual machine is an abstract unit of storage and compute capacities provided in a cloud. Without loss of generality, we assume that VMs from different clouds are offered in different types, each of which has different characteristics. For example, they may have different numbers of CPUs, amounts of memory, and network bandwidths. As well, the computational characteristics of different CPUs may not be the same.

In our proposed cloud resource allocation mechanism, every provider has a scheduler software running in its data center. These schedulers know the current statuses of VMs in their own clouds. And the schedulers communicate with each other, so they can make schedule decisions when using the information like the earliest resource available time in a certain cloud. Due to the security issues, we limit the types of communication functions among schedulers as follows. The scheduler of cloud A can send a task check request to the scheduler of cloud B. The scheduler of B respond with the earliest available time of all required resources, based on the current status of resources. When B finishes a task which was transferred from the scheduler of A, the scheduler of B will inform the scheduler of A that the task is finished.

When a job is submitted to a cloud, the scheduler first partitions the job into several tasks. Then for each task in this job, the scheduler decides which cloud will execute this task based on the information from all other schedulers. If the scheduler assigns a task to its own cloud, it will store the task in a queue. And when the resources and the data are ready, this task's execution begins. If the scheduler of cloud A assigns a task to cloud B, the scheduler of B first checks whether its resource availabilities can meet the requirement of this task. If so, the task will enter a queue waiting for execution. Otherwise, the scheduler of B will reject the task.

Before a task in the queue of a scheduler is about to be executed, the scheduler transfers a disk image to all the computing nodes which provide enough VMs for task execution. We assume that all required disk images are stored in a data center from which the images can be transferred to any clouds as needed. Assuming the size of this disk image is S_I, we use the multicasting and model the transfer time as S_I/b. b is the network bandwidth. When a VM finishes its part of the task, the disk image is discarded.

10.3.2 Resource Allocation Model

In cloud computing, there are three different modes of renting the computing capacities from a cloud provider:

- Advance Reservation (AR): Resources are reserved in advance. They should be available at a specific time.
- Best-Effort: Resources are provisioned as soon as possible. Requests are placed in a queue.
- Immediate: When a client submits a request, either the resources are provisioned immediately, or the request is rejected, based on the resource availabilities.

A lease of resource is implemented as a set of VMs. And the allocated resources of a lease can be described by a tuple (n, m, d, b), where n is number of CPUs, m is memory in megabytes, d is disk space in megabytes, and b is the network bandwidth in megabytes per second. For the AR mode, the lease also includes the required start time and the required execution time. For the best-effort and the immediate modes, the lease has information about how long the execution lasts, but not the start time of execution.

As shown in [193], combining AR and best-effort in a preemptable fashion can overcome the utilization problems. In this chapter, we assume that a few jobs submitted in the cloud system are in the AR mode, while the rest of the jobs are in the best-effort mode. And the jobs in AR mode have higher priorities, and are able to preempt the executions of the best-effort jobs.

When an AR task A needs to preempt a best-effort task B, the VMs have to suspend task B and restore the current disk image of task B in a specific disk space before the scheduler transfers the disk image of tasks A to the VMs. Assuming the speed of copy data in the disk is S_c, and the size of B's disk image is M_B, then the preparation overhead is M_B/S_c. When the task A finishes, the VMs will resume the execution of task B. Similar to the preparation overhead, the reloading overhead, which is caused by reloading the disk image of task B back, is also M_B/S_c. We assume that there is a specific disk space in every node for storing the disk image of the suspended task.

10.3.3 Job Model

In this chapter, we use the *Directed Acyclic Graphs* (DAG) to represent jobs. A DAG $T = (V, E)$ consists of a set of vertices V, each of which represents a task in the job, and a set of edges E, showing the dependencies among the tasks. The edge set E contains edges e_{ij} for each task $v_i \in V$ that task $v_j \in V$ depends on. The weight of a task represents the task type of this task. Given an edge e_{ij}, v_i is the immediate predecessor of v_j, and v_j is called the immediate successor of v_i. A task only starts after all its immediate predecessors finish. Tasks with no immediate predecessor are entry-tasks, and tasks without immediate successors are exit-tasks.

Although the compute nodes from the same cloud may be equipped with different hardware, the scheduler can treat its cloud as a homogeneous system by using the abstract compute capacity unit, virtual machine. However, as we assume, the VMs from a different cloud may have different characteristics. So the whole cloud system is a heterogeneous system. In order to describe the difference between VMs' computational characteristics, we use an $M \times N$ execution time matrix (ETM); E to indicate the execution time of M types of tasks running on N types of VMs. For example, the entry e_{ij} in E indicates the required execution time of task type i when running on VM type j. We also assume that a task requires the same lease (n, m, d, b) no matter on which type of VM the task is about to run.

10.4 Resource Allocation and Task Scheduling Algorithm

Since the schedulers neither know when jobs arrive, nor whether other schedulers receive jobs, it is a dynamic scheduling problem. We propose two algorithms for the task scheduling: *adaptive list scheduling* (ALS) and *adaptive min-min scheduling* (AMMS). Once a job is submitted to a scheduler, it will be partitioned into tasks in the form of DAG. In both ALS and AMMS, a static resource allocation, which includes a static task scheduling, is generated offline. The tasks are assigned to certain cloud resources based on the static resource allocation. Then the scheduler will repeatedly reevaluate the remaining static allocation with a predefined frequency, based on the latest information of task execution on other clouds. For a given task, if the expected finish time, which is calculated in the reevaluation, is later than the static estimated finish time by a predefined threshold, the scheduler will reallocate the resources for this task. More details are shown later in this section.

10.4.1 Static Resource Allocation

As we mentioned above, when a scheduler receives a job submission, it will first partition this job into tasks in the form of a DAG. Then a static resource allocation is generated offline. We proposed two greedy algorithms to generate the static allocation: the cloud list scheduling and the cloud min-min scheduling.

10.4.1.1 Cloud List Scheduling (CLS)

Our proposed CLS is similar to CPNT [83]. Some definitions used in listing the task are provided as follows. The *earliest start time* (EST) and the *latest start time* (LST) of a task are shown in Equations (10.1) and (10.2). The entry-tasks have EST equal to 0 and the LST of exit-tasks equal to their EST.

$$EST(v_i) = \max_{v_m \in pred(v_i)} \{EST(v_m) + AT(v_m)\} \tag{10.1}$$

$$LST(v_i) = \min_{v_m \in succv_i} \{LST(v_m)\} - AT(v_i) \tag{10.2}$$

Because the cloud system concerned in this chapter is heterogeneous, the execution times of a task on VMs of different clouds are not the same. $AT(v_i)$ is the average execution time of task v_i. The critical node (CN) is a set of vertices in the DAG of which EST and LST are equal. Algorithm 10.4.1 shows a function forming a task list based on the priorities.

Algorithm 10.4.1 Forming a task list based on the priorities

Require: A DAG, average execution time AT of every task in the DAG
Ensure: A list of tasks P based on priorities
1: The EST of every task is calculated.
2: The LST of every task is calculated.
3: Empty list P and stack S, and pull all tasks in the list of task U.
4: Push the CN task into stack S in the decreasing order of their LST.
5: **while** the stack S is not empty **do**
6:     **if** $top(S)$ has un-stacked immediate predecessors **then**
7:         $S \leftarrow$the immediate predecessor with least LST
8:     **else**
9:         $P \leftarrow top(S)$
10:        pop $top(S)$
11:    **end if**
12: **end while**

Once the list of tasks is formed, we can allocate resources to tasks in the order of this list. The task on the top of this list will be assigned to the cloud that can finish it at the earliest time. Note that the task being assigned at this moment will start execution only when all its predecessor tasks are finished and the cloud resources allocated to it are available. After being assigned, this task is removed from the list. The procedure repeats until the list is empty. A static resource allocation is obtained after this assigning procedure which is shown in Algorithm 10.4.2.

10.4.1.2 Cloud Min-Min Scheduling (CMMS)

Min-min is another popular greedy algorithm [94]. The original min-min algorithm does not consider the dependencies among tasks. So in the dynamic

Algorithm 10.4.2 The assigning procedure of CLS

Require: A priority-based list of tasks P, m different clouds, ETM matrix
Ensure: A static resource allocation generated by CLS
1: **while** the list P is not empty **do**
2: $T = top(P)$
3: Send task check requests of T to all other schedulers
4: Receive the earliest resource available time responses for T from all other schedulers
5: Find the cloud C_{min} giving the earliest estimated finish time of T, assuming no other task preempts T
6: Assign task T to cloud C_{min}.
7: Remove T from P
8: **end while**

min-min algorithm used in this chapter, we need to update the mappable task set in every scheduling step to maintain the task dependencies. Tasks in the mappable task set are the tasks whose predecessor tasks are all assigned. Algorithm 10.4.3 shows the pseudo codes of the CMMS algorithm.

Algorithm 10.4.3 Cloud min-min scheduling (CMMS)

Require: A set of tasks, m different clouds, ETM matrix
Ensure: A schedule generated by CMMS
1: form a mappable task set P
2: **while** there are tasks not assigned **do**
3: update mappable task set P
4: **for** i: task $v_i \in P$ **do**
5: Send task check requests of v_i to all other schedulers
6: Receive the earliest resource available time responses from all other schedulers
7: Find the cloud $C_{min}(v_i)$ giving the earliest finish time of v_i, assuming no other task preempts v_i
8: **end for**
9: Find the task-cloud pair$(v_k, C_{min}(v_k))$ with the earliest finish time in the pairs generated in for-loop
10: Assign task v_k to cloud $D_{min}(v_k)$
11: Remove v_k from P
12: update the mappable task set P
13: **end while**

10.4.1.3 Local Resource Allocation

A scheduler uses a slot table to record execution schedules of all resources, i.e., VMs, in its cloud. When an AR task is assigned to a cloud, the scheduler

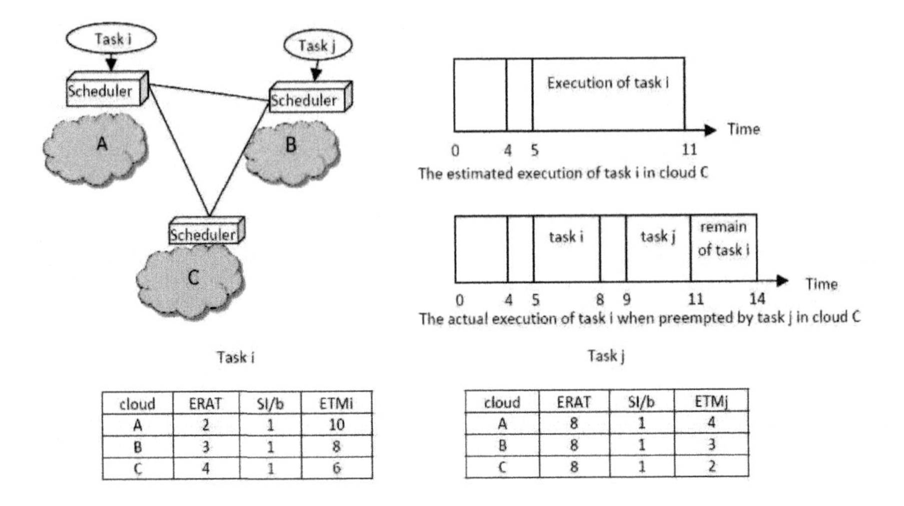

cloud	ERAT	SI/b	ETMi
A	2	1	10
B	3	1	8
C	4	1	6

cloud	ERAT	SI/b	ETMj
A	8	1	4
B	8	1	3
C	8	1	2

FIGURE 10.1: Example of resource contention.

of this cloud will first check the resource availability in this cloud. Since AR tasks can preempt best-effort tasks, the only case where an AR task is rejected is when most of the resources are reserved by some other AR tasks, and there are not enough resources left for this task in the required time slot. If the AR task is not rejected, which means there are enough resources for the task, a set of required VMs are selected arbitrarily. The time slots for transfer disk image of the AR task and the task execution are reserved in the slot tables of those VMs. The time slots for storing and reloading the disk image of the preempted task are also reserved if preemption happens.

When a best-effort task arrives, the scheduler will put it in the execution queue. Every time when there are enough VMs for the task on the top of the queue, a set of required VMs are selected arbitrarily for executing this best-effort task. And the scheduler also updates the time slot table of those VMs.

10.4.2 Online Adaptive Scheduling

Due to the resource contention within an individual cloud, the actual finish time of a task may not be the same as the estimated. We propose an online adaptive scheduling procedure to adjust the resource allocation dynamically based on the latest information.

But the estimated finish time from Equation (10.3) may not be accurate. For example, as shown in Figure 10.1, we assume three clouds in the system. The scheduler of cloud A needs to assign a best-effort task i to a cloud. According to Equation (10.3), cloud C has the smallest τ. So scheduler A transfers task i to cloud C. Then scheduler of cloud B needs to assign an AR task j to a cloud. Task j needs to reserve the resource at 8. Cloud C has the smallest τ again. Scheduler B transfers task j to cloud C. Since task j needs to start when i is not done, task j preempts task i at time 8. In this case, the actual finish time of task i is not the same as expected.

$$\tau_{i,j} = ERAT_{i,j} + S_I/b + ETM_{i,j} \tag{10.3}$$

In order to reduce the impacts of this kind of delay, we use an online adaptive procedure to adjust the schedules.

Our proposed online adaptive procedure will reevaluate the remaining static resource allocation repeatedly with a predefined frequency. In each reevaluation, the schedulers will recalculate the estimated finish time of their tasks. Note that a scheduler of a given cloud will only reevaluate the tasks that are in the jobs submitted to this cloud, not the tasks that are assigned to this cloud. A new estimated earliest finish time $\tau_{fdi,j}$ of task i running on cloud j, based on the latest information, is as follows:

$$\tau_{fd_{i,j}} = ERAT\_updated_{i,j} + S_I/b + ETM_{i,j} \tag{10.4}$$

where $ERAT\_updated_{i,j}$ is the updated earliest resource available time.

And we further calculate the difference between the original estimated finish time and the new one:

$$D_{\tau_{i,j}} = \tau_{fd_{i,j}} - \tau_{d_{i,j}} \tag{10.5}$$

For the tasks with $D_{\tau_{i,j}} > D_{threshold}$, where $D_{threshold}$ is a threshold, the scheduler will reschedule these tasks with a predefined probability α. These tasks will be reassigned to the clouds that give the minimum $\tau_{fd_{i,j}}$. We assume the scheduler will reevaluate the remaining static resource allocation before the next task k, which is assigned on cloud C, is executed. The $\tau_{fd_{k,C}}$ is computed with the latest information and the task k will be reassigned if the $D_{\tau_{k,C}}$ is larger than $D_{threshold}$.

10.5 Experimental Results

10.5.1 Experiment Setup

We evaluate the performance of our adaptive algorithms through simulations. In each run of our simulation, we simulate a set of 64 different jobs, and each job is composed of up to 16 tasks. There are four clouds in the simulation. Those 64 jobs will be submitted to random clouds at arbitrary arrival time. Among these 64 jobs, 12 jobs are in the AR modes, while the rest are in the best-effort modes. We set parameters in the simulation randomly according to the maximum and minimum values shown in Table 10.1. The whole simulation includes 10 runs with different sets of jobs. Since we focus on the scheduling algorithms, we do our simulations locally without implementing in any exiting cloud system or using VM interface API.

parameter	Minimum	Maximum
$ETM_{i,j}$	25	100
number of VMs in a cloud	20	100
number of CPU in a VM	1	8
network bandwidth in a VM	2	100
memory in a VM	32	2048
disk space in a VM	5,000	100,000
number of CPU in a lease	50	200
network bandwidth in a lease	10	800
memory in a lease	100	10000
disk space in a lease	500,000	20,000,000
Speed of copy in disk	100	1000

TABLE 10.1: Ranges of model parameters.

We set the arrival of jobs in two different ways. In the first way, we spread the arrival time of jobs widely over time so that, in most of the cases, jobs do not need to contend resources in cloud. We call this loose situation. In the other way, we set the arrival time of jobs close to each other. It means that jobs usually need to wait for resources in cloud. We call this tight situation. In both these two settings, we tune the constant α to show how the adaptive procedure impacts the average job execution time. We define the execution time of a job as the time elapsed from when the job is submitted to when the job is finished.

10.5.2 Result

Figure 10.2 shows the average job execution time in the loose situation. We find out that the AMMS algorithm has the shorter average execution time. And the adaptive procedure with updated information does not impact the job execution time significantly. The reason the adaptive procedure does not have a significant impact on the job execution time is that the resource contention is not significant in the loose situation. Most of the resource contention occurs when an AR job preempts a best-effort job. So the estimated finish time of a job is usually close to the actual finish time, which limits the effect of the adaptive procedure. And the scheduler does not call the adaptive procedure in most of the cases.

Figure 10.3 shows that AMMS still outperforms ALS. And the adaptive

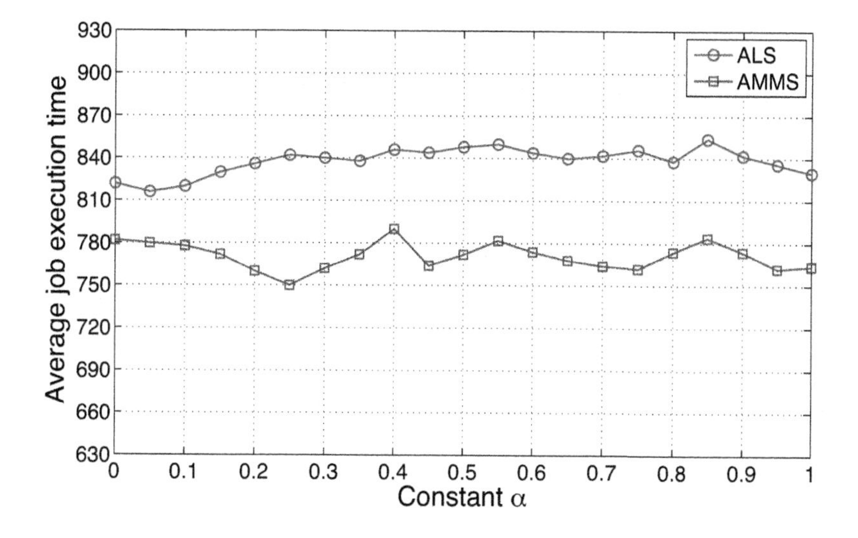

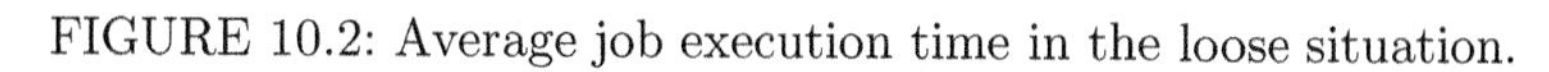
FIGURE 10.2: Average job execution time in the loose situation.

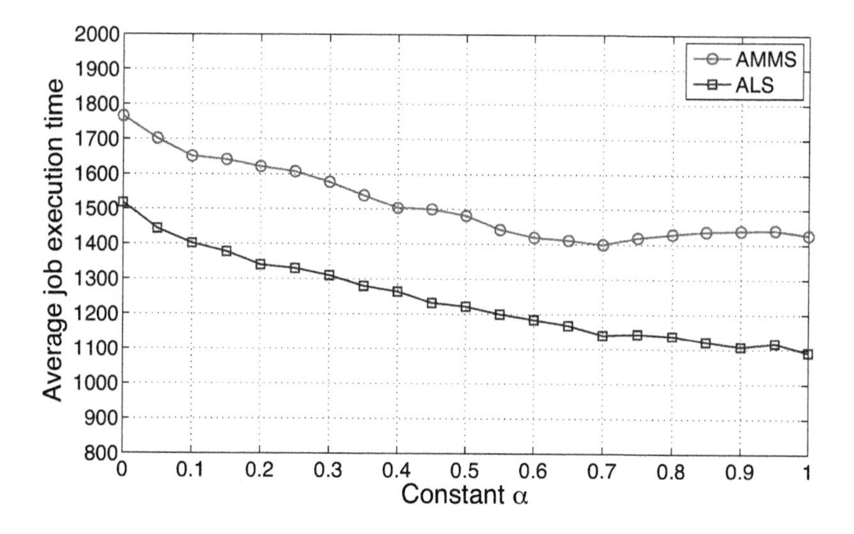

FIGURE 10.3: Average job execution time in the tight situation.

procedure with updated information works more significantly in the tight situation than it does in the loose situation. Because the resource contentions are fiercer in tight situation, the actual finish time of a task is often later than estimated finish time. And the best-effort task is more likely preempted by some AR tasks. The adaptive procedure can avoid tasks gathering in some fast clouds. We believe that the adaptive procedure works even better in a homogeneous cloud system, in which every task runs faster in some kinds of VMs than in some other kinds.

10.6 Conclusion

In this chapter, we present a resource allocation mechanism for preemptable jobs in cloud systems. And we propose two adaptive scheduling algorithms for this resource allocation mechanism. Simulation results show that the AMMS outperforms ALS. And the adaptive procedure with updated information provides significant improvement in the fierce resource contention situation.

10.7 Glossary

Cloud: In cloud computing, a cloud is a cluster of distributed computers providing on-demand computational resources or services to the remote users over a network.

IaaS: In an Infrastructure-as-a-Service (IaaS) cloud, resources or services are provided to users in the form of leases.

Virtual Machines: A software implementation of a machine that executes instructions like a physical machine.

Bibliography

[1] Cpu12 reference manual. *http://www.motorola.com/brdata/pdfdb/microcontrollers/16bit/68hc12family/refmat/cpu12rm.pdf.*

[2] Failure mechanisms and models for semiconductor devices. *JEDEC publication, http://www.jedec.org.*

[3] Mmc2001 reference manual. *http://www.motorola.com/SPS/MCORE/info_documentation.htm.*

[4] Tms370cx7x 8-bit microcontroller. *http://www-s.ti.com/sc/psheets/spns034c/spns034c.pdf.*

[5] Valgrind. *http://www.valgrind.org.*

[6] A. Aiken and A. Nicolau. Optimal loop parallelization. *SIGPLAN Not.*, 23(7), 1988.

[7] I.F. Akyildiz, Y. Sankarasubramaniam, W. Su, and E. Cayirci. A survey on sensor networks. *IEEE Communications Magazine*, 40(8):102–116, Aug. 2002.

[8] I.F. Akyildiz, W. Su, Y. Sankarasubramaniam, and E. Cayirci. Wireless sensor networks: a survey. *Computer Networks*, 38(4):393–422, 2002.

[9] J. Al-Jaroodi, N. Mohamed, H. Jiang, and D. Swanson. Modeling parallel applications performance on heterogeneous systems. In *Proc. IEEE IPDPS, Workshop Advances in Parallel and Distributed Computational Models*, 2003.

[10] S. Ali, A.A. Maciejewski, H.J. Siegel, and J.K. Kim. Measuring the robustness of a resource allocation. *IEEE Transactions on Parallel and Distributed Systems*, pages 630–641, 2004.

[11] S. Ali, H.J. Siegel, M. Maheswaran, D. Hensgen, and S. Ali. Representing task and machine heterogeneities for heterogeneous computing systems. *Tamkang Journal of Science and Engineering*, 3(3):195–208, 2000.

[12] N. Allec, Z. Hassan, L. Shang, R.P. Dick, and R. Yang. Thermalscope: Multi-scale thermal analysis for nanometer-scale integrated circuits. In *ACM/IEEE ICCAD*, pages 75–82, 2008.

[13] Analog Devices, Inc. *ADSP-21 000 Family Application Handbook Volume 1*, Norwood, MA, 1994.

[14] R. Armstrong, D. Hensgen, and T. Kidd. The relative performance of various mapping algorithms is independent of sizable variances in run-time predictions. In *7th IEEE Heterogeneous Computing Workshop*, volume 5, 1998.

[15] O. Avissar, R. Barua, and D. Stewart. Heterogeneous memory management for embedded systems. In *CASES*, pages 34–43, 2001.

[16] H. Aydin, R. Melhem, D. Mosse, and P. Alvarez. Dynamic and aggressive scheduling techniques for power aware real-time systems. In *RTSS*, 2001.

[17] R. Ayoub and T.S. Rosing. Predict and act: dynamic thermal management for multi-core processors. In *Proc. ACM ISLPED*, pages 99–104, San Francisco, CA, USA, 2009.

[18] R. Banakar, S. Steinke, B.-S. Lee, M. Balakrishnan, and P. Marwedel. Scratchpad memory: A design alternative for cache on-chip memory in embedded systems. *CODES '02: Proceedings of the tenth international symposium on Hardware/software codesign*, pages 73–78, 2002.

[19] C. Banino, O. Beaumont, L. Carter, J. Ferrante, A. Legrand, and Y. Robert. Scheduling strategies for master-slave tasking on heterogeneous processor platforms. *IEEE Trans. on Parallel Distributed Systems*, 15(4):319–330, 2004.

[20] P. Basu, W. Ke, and T.D.C. Little. Dynamic task-based anycasting in mobile ad hoc networks. *Mobile Networks and Applications*, 8(5):593–612, 2003.

[21] O. Beaumont, A. Legrand, L. Marchal, and Y. Robert. Scheduling strategies for mixed data and task parallelism on heterogeneous clusters. *Parallel Processing Letters*, 13(2):225–244, 2003.

[22] O. Beaumont, A. Legrand, L. Marchal, and Y. Robert. Pipelining broadcasts on heterogeneous platforms. In *International Parallel and Distributed Processing Symposium IPDPS'2004*. IEEE Computer Society Press, 2004.

[23] O. Beaumont, A. Legrand, and Y. Robert. Static scheduling strategies for heterogeneous systems. *Computing and Informatics*, 21:413–430, 2002.

[24] O. Beaumont, A. Legrand, and Y. Robert. The master-slave paradigm with heterogeneous processors. *IEEE Trans. on Parallel Distributed Systems*, 14(9):897–908, 2003.

[25] P. Berman, G. Calinescu, C. Shah, and A. Zelikovsly. Efficient energy management in sensor networks. *Ad Hoc and Sensor Networks*, 2005.

[26] G. Bernat, A. Colin, and S.M. Petters. WCET analysis of probabilistic hard real-time systems. In *IEEE Real-Time Systems Symposium*, pages 279–288, 2002.

[27] R. Bettati and J.W.-S. Liu. End-to-end scheduling to meet deadlines in distributed systems. In *Proc. of the International Conf. on Distributed Computing Systems*, pages 452–459, Jun. 1992.

[28] R. Bianchini, R. Pinto, and C.L. Amorim. Data prefetching for software dsms. In *Int. Conf. Supercomput.*, Jul. 1998.

[29] B. Black, M. Annavaram, N. Brekelbaum, J. DeVale, L. Jiang, G.H. Loh, D. McCaule, P. Morrow, D.W. Nelson, and D. Pantuso. Die stacking (3D) microarchitecture. In *Proceedings of the 39th Annual IEEE/ACM International Symposium on Microarchitecture*, pages 469–479, 2006.

[30] J. Bolot and A. Vega-Garcia. Control mechanisms for packet audio in the internet. In *Proceedings of IEEE Infocom*, 1996.

[31] P. Bouilet, A. Darte, T. Risset, and Y. Robert. (pen)-ultimate tiling. *Integration, VLSI J.*, 17, 1994.

[32] T.D. Braun, H.J. Siegel, N. Beck, L.L. Boloni, M. Maheswaran, A.I. Reuther, J.P. Robertson, M.D. Theys, B. Yao, and D. Hensgen. A comparison of eleven static heuristics for mapping a class of independent tasks onto heterogeneous distributed computing systems. *Journal of Parallel and Distributed Computing*, 61(6):810–837, 2001.

[33] T.D. Braun, H.J. Siegel, A.A. Maciejewski, and S.D. Noemix. Static mapping heuristics for tasks with dependencies, priorities, deadlines, and multiple versions in heterogeneous environments. In *International Parallel and Distributed Processing Symposium*, pages 78–85, 2002.

[34] P. Brereton and D. Budgen. Component-based systems: A classification of issues. *Computer*, 33(11):54–62, Nov. 2000.

[35] D. Brooks, V. Tiwari, and M. Martonosi. Wattch: A framework for architectural-level power analysis and optimizations. In *IEEE ISCA*, pages 83–94, 2000.

[36] R. Buyya. High Performance Cluster Computing: Architectures and Systems. Upper Saddle River, NJ: Prentice-Hall PTR, 1999.

[37] A. Cerpa and D. Estrin. Ascent: Adaptive self configuring sensor networks topologies. In *Proceedings of IEEE INFOCOM2002, New York, NY*, Jun. 2002.

[38] W. Cesário, A. Baghdadi, D. Lyonnard, G. Nicolescu, Y. Paviot, S. Yoo, A.A. Jerraya, and M. Diaz-Nava. Component-based design approach for multicore socs. In *Proc. of the Design Automation Conf.*, Jun. 2002.

[39] J. Chame and S. Moon. A tile selection algorithm for data locality and cache interference. In *ACM Int. Conf. Supercomput.*, pages 492–499, Rhodes, Greece, Jun. 1999.

[40] Y.-N. Chang, C.-Y. Wang, and K.K. Parhi. Loop-list scheduling for heterogeneous functional units. In *6th Great Lakes Symposium on VLSI*, pages 2–7, Mar. 1996.

[41] L.-F. Chao, A. LaPaugh, and E.H.-M. Sha. Rotation scheduling: A loop pipelining algorithm. *IEEE Trans. on Computer-Aided Design of Integrated Circuits and Systems*, 16:229–239, Mar. 1997.

[42] L.-F. Chao, A.S. LaPaugh, and E.H.-M. Sha. Rotation scheduling: A loop pipelining algorithm. *IEEE Transactions on Computer-Aided Design*, 16(3):229–239, Mar. 1997.

[43] L.-F. Chao and E.H.-M. Sha. Static scheduling for synthesis of DSP algorithms on various models. *Journal of VLSI Signal Processing Systems*, 10:207–223, 1995.

[44] L.-F. Chao and E.H.-M. Sha. Scheduling data-flow graphs via retiming and unfolding. *IEEE Trans. on Parallel and Distributed Systems*, 8:1259–1267, Dec. 1997.

[45] L.-F. Chao and E.H.-M. Sha. Scheduling data-flow graphs via retiming and unfolding. *IEEE Transactions on Parallel and Distributed Systems*, 8(12):1259–1267, Dec. 1997.

[46] I. Chatzigiannakis, A. Kinalis, and S. Nikoletseas. Power conservation schemes for energy efficient data propagation in heterogeneous wireless sensor networks. In *Proceedings of the 38th Annual Symposium on Simulation*, pages 60–71, Apr. 2005.

[47] A. Chavez, C. Tornabene, and G. Wiederhold. Software component licensing: A primer. *IEEE Software*, pages 47–53, 1998.

[48] F. Chen, T.W. O'Neil, and E.H.-M. Sha. Optimizing overall loop schedules using prefetching and partitioning. *IEEE Trans. Parallel Distrib. Syst.*, 11(6), 2000.

[49] F. Chen and E.H.-M. Sha. Loop scheduling and partitions for hiding memory latencies. In *ISSS*, pages 64–70, 1999.

[50] J.-J. Chen and T.-W. Kuo. Multiprocessor energy-efficient scheduling for real-time tasks with different power characteristics. In *ICPP*, 2005.

[51] T.-F. Chen and J.-L. Baer. A performance study of software and hardware data prefetching schemes. In *Ann. Int. Symp. Comput. Architecture*, pages 223–232, 1994.

[52] T.F. Chen and J.L. Baer. Effective hardware-based prefetching for high-performance microprocessors. *IEEE Transactions on Computers*, pages 609–623, May 1995.

[53] Y. Chen, Z. Shao, Q. Zhuge, C. Xue, B. Xiao, and E.H.-M. Sha. Minimizing energy via loop scheduling and DVS for multi-core embedded systems. In *ICPADS'05 Volume II and PDES 2005*, pages 2–6, Fukuoka, Japan, 20-22 Jul. 2005.

[54] J. Cho, Y. Paek, and D. Whalley. Efficient register and memory assignment for non-orthogonal architectures via graph coloring and mst algorithms. In *ACM Joint Conference LCTES-SCOPES*, pages 130–138, Berlin, Germany, Jun. 2002.

[55] P. Chou and G. Borriello. An analysis-based approach to composition of distributed embedded systems. In *CODES98*, Seattle, WA, 1998.

[56] E.G. Coffman and J.L. Bruno. *Computer and Job-Shop Scheduling Theory.* New York: John Wiley & Sons, 1976.

[57] A.K. Coskun, J.L. Ayala, D. Atienza, T.S. Rosing, and Y. Leblebici. Dynamic thermal management in 3D multicore architectures. In *ACM/IEEE DATE*, pages 1410–1415, 2009.

[58] F. Dahlgren and M. Dubois. Sequential hardware prefetching in shared-memory multiprocessors. *IEEE Trans. Parallel Distrib. Syst.*, 6, Jul. 1995.

[59] L. David and I. Puaut. Static determination of probabilistic execution times. In *Euromicro Conference on Real-Time Systems*, pages 223–230.

[60] V. Delaluz, M. Kandemir, and I. Kolcu. Automatic data migration for reducing energy consumption in multi-bank memory systems. In *DAC*, pages 213–218, New Orleans, LA, USA 2002.

[61] V. Delaluz, M. Kandemir, A. Sivasubramaniam, and M.J. Irwin. Hardware and software techniques for controlling dram power modes. *IEEE Trans. on Computers*, 50(11), Nov. 2001.

[62] J. Deng, Y.S. Han, W.B. Heinzelman, and P.K. Varshney. Balanced-energy sleep scheduling scheme for high density cluster-based sensor networks. *Elsevier Computer Communications Journal, Special Issue on ASWN '04*, 2004.

[63] J. Deng, Y.S. Han, W.B. Heinzelman, and P.K. Varshney. Scheduling sleeping nodes in high density cluster based sensor networks. *ACM/Kluwer Mobile Networks and Applications (MONET) Special Issue on Energy Constraints and Lifetime Performance in Wireless Sensor Networks*, 2004.

[64] A. Dogan and F. Özgüner. Matching and scheduling algorithms for minimizing execution time and failure probability of applications in heterogeneous computing. *IEEE Trans. on Parallel and Distributed Systems*, 13:308–323, Mar. 2002.

[65] A. Dogan and F. Ozguner. Matching and scheduling algorithms for minimizing execution time and failure probability of applications in heterogeneous computing. *IEEE Transactions on Parallel and Distributed Systems*, pages 308–323, 2002.

[66] A. Easwaran, I. Shin, O. Sokolsky, and I. Lee. Incremental schedulability analysis of hierarchical real-time components. In *EMSOFT'06, Seoul, Korea.*, October 22–25, 2006.

[67] J. Elson and D. Estrin. Time synchronization for wireless sensor networks. In *Proceedings of the 15th International Parallel and Distributed Processing Symposium (IPDPS '01)*, 2001.

[68] W. Emeneker and D. Stanzione. Efficient Virtual Machine Caching in Dynamic Virtual Clusters. In *SRMPDS Workshop, ICAPDS*, 2007.

[69] M.M. Eshaghian. *Heterogeneous computing.* Artech House, 1996.

[70] N. Fallenbeck, H.J. Picht, M. Smith, and B. Freisleben. Xen and the art of cluster scheduling. In *Proceedings of the 2nd International Workshop on Virtualization Technology in Distributed Computing*, page 4, 2006.

[71] W. Feng, E. Kaiser, W.C. Feng, and M. Le Baillif. Panoptes: scalable low-power video sensor networking technologies. *ACM Transactions on Multimedia Computing, Communications, and Applications (TOMCCAP)*, 1(2):151–167, 2005.

[72] D. Fernandez-Baca. Allocating modules to processors in a distributed system. *IEEE Transactions on Software Engineering*, 15(11):1427–1436, 1989.

[73] I. Foster. *Designing and Building Parallel Program: Concepts and Tools for Parallel Software Engineering.* Boston: Addison-Wesley, 1994.

[74] R.F. Freund, M. Gherrity, S. Ambrosius, M. Campbell, M. Halderman, D. Hensgen, E. Keith, T. Kidd, M. Kussow, J.D. Lima, F. Mirabile, L. Moore, B. Rust, and H.J. Siegel. Scheduling resources in multi-user, heterogeneous, computing environments with SmartNet. In *IEEE Heterogeneous Computing Workshop*, pages 184–199, 1998.

[75] J.W.C. Fu and J.H. Patel. Stride directed prefetching in scalar processors. In *25th Intl. Symp. Microarchitecture*, pages 102–110, Dec. 1992.

[76] M.R. Garey and D.S. Johnson. *Computers and Intractability: A Guide to the Theory of NP-Completeness.* San Francisco, CA: W. H. Freeman, 1979.

[77] C.H. Gebotys and M. Elmasry. Global optimization approach for architectural synthesis. *IEEE Trans. on Computer-Aided Design of Integrated Circuits and Systems*, 12:1266–1278, Sep. 1993.

[78] T. Genbler, O. Nierstrasz, and B. Schonhage. Components for embedded software. In *CASES 2002, Grenoble, France*, October 8–11, 2002.

[79] C. Germain-Renaud and O.F. Rana. The convergence of clouds, grids, and autonomics. *IEEE Internet Computing*, page 9, 2009.

[80] S. Giannecchini, M. Caccamo, and C.S. Shih. Collaborative resource allocation in wireless sensor networks. In *Proceedings of the 16th Euromicro Conference on Real-Time Systems*, pages 35–44.

[81] M. Gschwind, H.P. Hofstee, B. Flachs, M. Hopkins, Y. Watanabe, and T. Yamazaki. Synergistic processing in cell's multicore architecture. *IEEE Micro*, pages 10–24, Mar.-Apr. 2006.

[82] M.R. Guthaus, J.S. Ringenberg, D. Ernst, T.M. Austin, T. Mudge, and R.B. Brown. Mibench: A free, commercially representative embedded benchmark suite. *WWC '01: Proceedings of the Workload Characterization, 2001. WWC-4. 2001 IEEE International Workshop*, pages 3–14, 2001.

[83] T. Hagras and J. Janecek. A high performance, low complexity algorithm for compile-time job scheduling in homogeneous computing environments. In *Parallel Processing Workshops, 2003. Proceedings. 2003 International Conference on Parallel Processing Workshops*, pages 149–155, 2003.

[84] Y. Han, I. Koren, and C.A. Moritz. Temperature aware floorplanning. In *Workshop on Temperature-Aware Computer Systems*, 2005.

[85] Y. He, Z. Shao, B. Xiao, Q. Zhuge, and E.H.-M. Sha. Reliability driven task scheduling for tightly coupled heterogeneous systems. In *Proc. of IASTED International Conference on Parallel and Distributed Computing and Systems*, Nov. 2003.

[86] P. Hofstee and M. Day. Hardware and software architectures for the cell processor. In *CODES+ISSS'05, Jersey City, New Jersey, USA.*, Sept. 19–21, 2005.

[87] K. Hogstedt, D. Kimelman, V. Rajan, T. Roth, M. Wegman, and N. Wang. Optimizing component interaction. In *OM 2001, Snowbird, Utah, USA*, 2001.

[88] C.-J. Hou and K.G. Shin. Allocation of periodic task modules with precedence and deadline constraints in distributed real-time systems. In *IEEE Trans. on Computers*, volume 46, pages 1338–1356, Dec. 1997.

[89] http://asic.amsint.com/databooks/digital/gepard.htm. *GEPARD Family of Embedded Software Programmable DSP Core.*

[90] S. Hua and G. Qu. Approaching the maximum energy saving on embedded systems with multiple voltages. In *International Conference on Computer Aid Design (ICCAD)*, pages 26–29, 2003.

[91] S. Hua, G. Qu, and S.S. Bhattacharyya. Energy reduction techniques for multimedia applications with tolerance to deadline misses. In *ACM/IEEE Design Automation Conference (DAC)*, pages 131–136, 2003.

[92] S. Hua, G. Qu, and S.S. Bhattacharyya. Exploring the probabilistic design space of multimedia systems. In *IEEE International Workshop on Rapid System Prototyping*, pages 233–240, 2003.

[93] C.-T. Hwang, J.-H. Lee, and Y.-C. Hsu. A formal approach to the scheduling problem in high level synthesis. *IEEE Trans. on Computer-Aided Design of Integrated Circuits and Systems*, 10:464–475, Apr. 1991.

[94] O.H. Ibarra and C.E. Kim. Heuristic algorithms for scheduling independent tasks on nonidentical processors. *Journal of the ACM*, pages 280–289, 1977.

[95] C. Im, H. Kim, and S. Ha. Dynamic voltage scheduling technique for low-power multimedia applications using buffers. In *Proc. of ISLPED*, 2001.

[96] T. Ishihara and H. Yasuura. Voltage scheduling problem for dynamically variable voltage processor. In *ISLPED*, pages 197–202, 1998.

[97] K. Ito, L. Lucke, and K. Parhi. Ilp-based cost-optimal dsp synthesis with module selection and data format conversion. *IEEE Trans. on VLSI Systems*, 6:582–594, Dec. 1998.

[98] K. Ito and K. Parhi. Register minimization in cost-optimal synthesis of dsp architecture. In *Proc. of the IEEE VLSI Signal Processing Workshop*, Oct. 1995.

[99] R. Jejurikar, C. Pereira, and R. Gupta. Leakage aware dynamic voltage scaling for real-time embedded systems. In *DAC*, pages 275–280, 2004.

[100] A. Kalavade and P. Moghe. A tool for performance estimation of networked embedded end-systems. In *Proceedings of Design Automation Conference*, pages 257–262, Jun. 1998.

[101] S. Kallakuri and A. Doboli. Energy conscious online architecture adaptation for varying latency constraints in sensor network applications. In *CODES+ISSS'05*, pages 148–154, Jersey City, New Jersey, Sept. 19–21 2005.

[102] M. Kandemir. Impact of data transformations on memory bank locality. In *DATE*, volume 1, pages 506–511, Feb. 2004.

[103] M. Kandemir, J. Ramanujam, and A. Choudhury. Exploring shared scratch pad memory space in embedded multiprocessor system. *DAC '02: Proceedings of the 39th Conference on Design Automation*, pages 219–224, 2002.

[104] M. Kandemir, J. Ramanujam, J. Irwin, N. Vijaykrishnan, I. Kadayif, and A. Parikh. Dynamic management of scratch-pad memory space. *DAC '01: Proceedings of the 38th Conference on Design Automation*, pages 690–695, 2001.

[105] D. Karlsson, P. Eles, and Z. Peng. Formal verification in a component-based reuse methodology. In *ISSS'02, Kyoto, Japan*, October 2-4, 2002.

[106] R.E. Kessler. The Alpha 21264 microprocessor. *IEEE Micro*, 19(2):24–36, Mar./Apr. 1999.

[107] M. Kistler, M. Perrone, and F. Petrini. Cell multiprocessors communication network: Built for speed. *IEEE Micro*, pages 10–23, May-Jun. 2006.

[108] R. Kumar, M. Wolenetz, B. Agarwalla, J.S. Shin, P. Hutto, A. Paul, and U. Ramachandran. DFuse: a framework for distributed data fusion. In *Proceedings of the 1st International Conference on Embedded Networked Sensor Systems*, pages 114–125. ACM, New York, 2003.

[109] S. Kumar, T.H. Lai, and J. Balogh. On k-coverage in a mostly sleeping sensor network. In *Proceedings of the 10th Annual International Conference on Mobile Computing and Networking (Mobicom '04)*, pages 144–158, 2004.

[110] Y.W. Law, J. Doumen L. Hoesel, and P. Havinga. Sensor networks: Energy-efficient link-layer jamming attacks against wireless sensor network mac protocols. In *Proceedings of the 3rd ACM Workshop on Security of Ad Hoc and Sensor Networks SASN '05*, Alexandria, VA, pages 76–88, Nov. 2005.

[111] C. Leangsuksun, J. Potter, and S. Scott. Dynamic task mapping algorithms for a distributed heterogeneous computing environment. In *4th IEEE Heterogeneous Computing Workshop*, pages 30–34, 1995.

[112] A.R. Lebeck, X. Fan, H. Zeng, and C.S. Ellis. Power aware page allocation. In *9th International Conference on Architectural Support for Programming Languages and Operating Systems*, Nov. 2000.

[113] B.C. Lee, E. Ipek, O. Mutlu, and D. Burger. Architecting phase change memory as a scalable DRAM alternative. In *Intl. Symp. on Computer Architecture (ISCA)*, 2009.

[114] B.C. Lee, P. Zhou, J. Yang, Y. Zhang, B. Zhao, E. Ipek, O. Mutlu, and D. Burger. Phase-change technology and the future of main memory. *IEEE Micro*, 30(1), 2010.

[115] C.E. Leiserson and J.B. Saxe. Retiming synchronous circuitry. *Algorithmica*, 6:5–35, 1991.

[116] B.P. Lester. *The Art of Parallel Programming.* Englewood Cliffs, NJ: Prentice Hall, 1993.

[117] R. Leupers and D. Kotte. Variable partitioning for dual memory bank dsps. In *IEEE Int. Conf. Acoust., Speech, Signal Process*, volume 2, pages 1121–1124, May 2001.

[118] M. Levy and T.M. Conte. Embedded Multicore Processors and Systems. *IEEE Micro*, 29(3):7–9, 2009.

[119] M. Li and F. Yao. An efficient algorithm for computing optimal discrete voltage schedules. *SIAM J. Comput.*, 35(3):658–671, 2005.

[120] W.N. Li, A. Lim, P. Agarwal, and S. Sahni. On the circuit implementation problem. *IEEE Trans. on Computer-Aided Design of Integrated Circuits and Systems*, 12:1147–1156, Aug. 1993.

[121] Y.A. Li, J.K. Antonio, H.J. Siegel, M. Tan, and D.W. Watson. Determining the execution time distribution for a data parallel program in a heterogeneous computing environment. *Journal of Parallel and Distributed Computing*, 44(1):35–52, 1997.

[122] S. Liu and M. Qiu. Thermal-aware scheduling for peak temperature reduction with stochastic workloads. In *IEEE/ACM RTAS*, Stockholm, Sweden, Apr. 2010.

[123] M. Lorenz, D. Kottmann, S. Bashfrod, R. Leupers, and P. Marwedel. Optimized address assignment for dsps with simd memory accesses. In *Asia South Pacific Design Automation Conference (ASP-DAC)*, pages 415–420, Yokohama, Japan, Jan. 2001.

[124] J. Luo and N. Jha. Static and dynamic variable voltage scheduling algorithms for real-time heterogeneous distributed embedded systems. In *VLSID*, 2002.

[125] D. Lyonnard, S. Yoo, A. Baghdadi, and A.A. Jerraya. Automatic generations of application-specific architectures for heterogeneous multiprocessor system-on-chip. In *Proc. of the Design Automation Conf.*, pages 518–523, Jun. 2001.

[126] C. Ma, Z. Zhang, and Y. Yang. Battery-aware router scheduling in wireless mesh networks. In *Proceedings of the 20th IEEE International Parallel and Distributed Processing Symposium (IPDPS'06), Rhodes Island, Greece*, 2006.

[127] G. Madl and S. Abdelwahed. Model-based analysis of distributed real-time embedded system composition. In *EMSOFT'05, Jersey City, New Jersey, USA.*, Sept. 19–22, 2005.

[128] M. Maheswaran, S. Ali, H.J. Siegel, D. Hensgen, and R.F. Freund. A comparison of dynamic strategies for mapping a class of independent tasks onto heterogeneous computing systems. In *Proc. of the Heterogeneous Computing Workshop*, pages 57–69, 1998.

[129] A. Mainwaring, J. Polastre, R. Szewczyk, D. Culler, and J. Anderson. Wireless sensor networks for habitat monitoring. In *Proc. of ACM International Worshop on Wireless Sensor Network Applications*, 2002.

[130] H.D. Man, F. Catthoor, G. Goossens, J. Vanhoof, J. Meerbergen, S. Note, and J.A. Huisken. Architecture-driven synthesis techniques for vlsi implementation of dsp algorithms. *Proceedings of the IEEE*, 78(2):319–335, 1990.

[131] N. Manjikian. Combining loop fusion with prefetching on shared-memory multiprocessors. In *Int. Conf. Parallel Process*, pages 78–82, 1997.

[132] M.C. McFarland, A.C. Parker, and R. Camposano. The high-level synthesis of digital systems. *Proceedings of the IEEE*, 78:301–318, Feb. 1990.

[133] S. Meftali, F. Gharsalli, F. Rousseau, and A.A. Jerraya. An optimal memory allocation for application-specific multiprocessor system-on-chip. *ISSS '01: Proceedings of the 14th International Symposium on Systems Synthesis*, pages 19–24, 2001.

[134] G.D. Micheli. *Synthesis and Optimization of Digital Circuits.* Burr Ridge, IL: McGraw-Hill, 1994.

[135] S.P. Mohanty and N. Ranganathan. Energy-efficient datapath scheduling using multiple voltages and dynamic clocking. *ACM Transactions on Design Automation of Electronic Systems (TODAES)*, 10(2):330–353, 2005.

[136] D. Mosse, H. Aydin, B. Childers, and R. Melhem. Compiler-assisted dynamic power-aware scheduling for real-time applications. In *Workshop on Compilers and Operating Systems for Low-Power*, 2000.

[137] Motorola. *DSP56000 24-Bit Digital Signal Processor Family Manual*, Schaumberg, IL, 1996.

[138] T. Mowry. Tolerating latency in multiprocessors through compiler-inserted prefetching. *ACM Trans. Comput. Syst.*, 16(1), 1998.

[139] G. Nicolescu, S. Yoo, A. Bouchhima, A.A. Jerraya, and M. Diaz-Nava. Validation in a component-based design flow for multicore socs. In *Proc. of the IEEE Int. Symp. on System Synthesis*, Oct. 2002.

[140] L. Niu and G. Quan. Reducing both dynamic and leakage energy consumption for hard real-time systems. In *Proceedings of the 2004 International Conference on Compilers, Architecture, and Synthesis for Embedded Systems (CASAS)*, pages 140–148, Washington DC, USA, Sept. 2004.

[141] G. Chen O. Ozturk, M. Kandemir, and M.J. Irwin. Multi-level on-chip memory hierachy design for embedded chip multiprocessor. *ICPADS '06: Proceedings of the 12th International Conference on Parallel and Distributed System*, pages 383–390, 2006.

[142] T. O'Neil and E.H.-M. Sha. Optimal graph transformation using extended retiming with minimal unfolding. In *Proceedings of the IASTED*, volume 4, pages 128–133, Nov. 2000.

[143] T. O'Neil, S. Tongsima, and E.H.-M. Sha. Extended retiming: Optimal scheduling via a graph-theoretical approach. In *Proc. of the IEEE Int. Conf. on Acoustics, Speech, and Signal Processing*, volume 4, pages 2001–2004, Mar. 1999.

[144] T. Ozawa, Y. Kimura, and S. Nishizaki. Cache miss heuristics and preloading techniques for general purpose programs. In *MICRO-28*, pages 243–248, 1995.

[145] O. Ozturk, G. Chen, M. Kandemir, and M. Karakoy. An integer linear programming based approach to simultaneous memory space partitioning and data allocation for chip multiprocessors. *ISVLSI '06: Proceedings of the IEEE Computer Society Annual Symposium on Emerging VLSI Technologies and Architectures*, page 50, 2006.

[146] O. Ozturk, M. Kandemir, G. Chen, M.J. Irwin, and M. Karakoy. Customized on-chip memories for embedded chip multiprocessors. *ASP-DAC '05: Proceedings of the 2005 conference on Asia South Pacific Design Automation*, pages 743–748, 2005.

[147] P.R. Panda, N.D. Dutt, and A. Nicolau. On-chip vs. off-chip memory: The data partitioning problem in embedded processor-based systems. *ACM Transactions on Design Automation of Embedded Systems*, 5(3), Jul. 2000.

[148] D. Panigrahi, C. Chiasserini, S. Dey, R. Rao, A. Raghunathan, and K. Lahiri. Battery life estimation of mobile embedded systems. In *Proceedings of the 14th International Conference on VLSI Design (VLSID'01)*. IEEE Computer Society, Washington, DC, USA, 2001.

[149] K. Parhi and D.G. Messerschmitt. Static rate-optimal scheduling of iterative data-flow programs via. optimum unfolding. *IEEE Trans. on Computers*, 40:178–195, Feb. 1991.

[150] A. Parikh, S. Kim, M. Kandemir, N. Vijaykrishnan, and M.J. Irwin. Instruction scheduling for low power. *Journal of VLSI Signal Processing*, 37:129–149, 2004.

[151] N. Passos and E.H.-M. Sha. Scheduling of uniform multi-dimensional systems under resource constraints. *IEEE Trans. VLSI Syst.*, 6, Dec. 1998.

[152] N. Passos and E.H.-M. Sha. Achieving full parallelism using multidimensional retiming. *IEEE Trans. Parallel and Distributed Systems*, 7(11), Nov. 1996.

[153] N. L. Passos, E.H.-M. Sha, and S.C. Bass. Loop pipelining for scheduling multi-dimensional systems via rotation. In *Proc. 31st Design Automation Conf.*, pages 485–490, Jun. 1994.

[154] M. Pathak and S. Lim. Thermal-aware steiner routing for 3D stacked ICs. In *ACM/IEEE ICCAD*, pages 205–211, 2008.

[155] P.G. Paulin and J.P. Knight. Force-directed scheduling for the behavioral synthesis of asic's. *IEEE Trans. on Computer-Aided Design of Integrated Circuits and Systems*, 8:661–679, Jun. 1989.

[156] P. Pillai and K.G. Shin. Real-time dynamic voltage scaling for low-power embedded operating systems. In *SOSP*, 2001.

[157] R. Pyne and E. Mugisa. Essential elements of a component-based development environment for the software supermarket. In *Proceedings of IEEE SouthEastern Conference*, North Carolina, 2004.

[158] M. Qiu, M. Guo, M. Liu, C. Xue, L.T. Yang, and E.H.-M. Sha. Loop scheduling and bank type assignment for heterogeneous multi-bank memory. *Journal of Parallel and Distributed Computing (JPDC)*, 69(5):546–558, May 2009.

[159] M. Qiu, Z. Jia, C. Xue, Z. Shao, and E.H.-M. Sha. Loop scheduling to minimize cost with data mining and prefetching for heterogeneous dsp. In *Proc. The 18th IASTED International Conference on Parallel and Distributed Computing Systems (PDCS 2006)*, Dallas, Texas, Nov. 13–15 2006.

[160] M. Qiu, Z. Jia, C. Xue, Z. Shao, and E.H.-M. Sha. Voltage assignment with guaranteed probability satisfying timing constraint for real-time multiproceesor dsp. *Journal of VLSI Signal Processing Systems for Signal, Image, and Video Technology (JVLSI)*, 2006.

[161] M. Qiu, M. Liu, C. Xue, Z. Shao, Q. Zhuge, and E.H.-M. Sha. Optimal assignment with guaranteed confidence probability for trees on heterogeneous dsp systems. In *Proceedings The 17th IASTED International Conference on Parallel and Distributed Computing Systems (PDCS 2005)*, Phoenix, Arizona, 14–16 Nov. 2005.

[162] M. Qiu and E.H.-M. Sha. Cost minimization while satisfying hard/soft timing constraints for heterogeneous embedded systems. *ACM Transactions on Design Automation of Electronic Systems (TODAES)*, 14(2, Article 25):1–30, Mar. 2009.

[163] M. Qiu and E.H. Sha. Cost minimization while satisfying hard/soft timing constraints for heterogeneous embedded systems. *ACM Transactions on Design Automation of Electronic Systems (TODAES)*, 14(2):1–30, 2009.

[164] M. Qiu, Z. Shao, C. Xue, Q. Zhuge, and E.H.-M. Sha. Heterogeneous assignment to minimize cost while satisfying hard/soft timing constraints. *Submitted to IEEE Trans. on Computer*, 2007.

[165] M. Qiu, Z. Shao, Q. Zhuge, C. Xue, M. Liu, and E.H.-M. Sha. Efficient assignment with guaranteed probability for heterogeneous parallel dsp. In *IEEE Int'l Conference on Parallel and Distributed Systems (ICPADS)*, pages 623–630, Minneapolis, MN, Jul. 2006.

[166] M. Qiu, C. Xue, Z. Shao, M. Liu, and E.H.-M. Sha. Energy minimization for heterogeneous wireless sensor networks. *Special Issue of Journal of Embedded Computing (JEC)*, 3(2):109–117, 2007.

[167] M. Qiu, L.T. Yang, Z. Shao, and E. Sha. Rotation scheduling and voltage assignment to minimize energy for SoC. In *International Conf. on Computational Science and Engineering*, pages 48–55, 2009.

[168] M. Qiu, L.T. Yang, Z. Shao, and E. Sha. Dynamic and leakage energy minimization with soft real-time loop scheduling and voltage assignment. *IEEE Transactions on Very Large Scale Integration Systems*, 18(3):501–504, 2010.

[169] M. Qiu, L. Zhang, and E.H.-M. Sha. ILP optimal scheduling for multimodule memory. In *ACM/IEEE CODES+ISSS 2009*, Grenoble, France, Oct. 2009.

[170] Q. Qiu, Q. Wu, and M. Pedram. Stochastic modeling of a power-managed system: construction and optimization. *IEEE Trans. on Computer Aided Design*, 20(10):1200–1217, Oct. 2001.

[171] M.K. Qureshi, V. Srinivasan, and J.A. Rivers. Scalable high performance main memory system using phase-change memory technology. In *Intl. Symp. on Computer Architecture (ISCA)*, 2009.

[172] M. Rahimi, R. Pon, W. Kaiser, G. Sukhatme, D. Estrin, and M. Srivastava. Adaptive sampling for environmental robots. In *International Conference on Robotics and Automation*, 2004.

[173] D. Rakhmatov and S. Vrudhula. Energy management for battery-powered embedded systems. *ACM Transactions on Embedded Computing Systems (TECS)*, 2(3):277–324, 2003.

[174] K. Ramamritham, J.A. Stankovic, and P.-F. Shiah. Efficient scheduling algorithms for real-time multiprocessor systems. In *IEEE Trans. on Parallel and Distributed Systems*, volume 1, pages 184–194, Apr. 1990.

[175] RAMBUS. *128/144-mbit direct RDRAM data sheet.* Rambus Inc., Los Altos, CA. Website: www.rambus.com., 1999.

[176] M. Renfors and Y. Neuvo. The maximum sampling rate of digital filters under hardware speed constraints. *IEEE Trans. on Circuits and Systems*, CAS-28:196–202, 1981.

[177] P. Ruth, P. McGachey, and D. Xu. Viocluster: Virtualization for dynamic computational domains. In *Proceedings of the IEEE International Conference on Cluster Computing*, pages 1–10, 2005.

[178] P. Ruth, J. Rhee, D. Xu, R. Kennell, and S. Goasguen. Autonomic live adaptation of virtual computational environments in a multi-domain infrastructure. In *IEEE International Conference on Autonomic Computing*, pages 5–14, 2006.

[179] M. Saghir, P. Chow, and C. Lee. Exploiting dual datamemory banks in digital signal processors. In *International Conference on Architecture Support for Programming Language and Operating Systems*, pages 234–243, 1996.

[180] H. Saputra, M. Kandemir, N. Vijaykrishnan, M.J. Irwin, J.S. Hu, C-H. Hsu, and U. Kremer. Energy-conscious compilation based on voltage scaling. In *LCTES'02*, Jun. 2002.

[181] Z. Shao, B. Xiao, C. Xue, Q. Zhuge, and E.H.-M. Sha. Loop scheduling with timing and switching-activity minimization for vliw dsp. *ACM Transactions on Design Automation of Electronic Systems (TODAES)*, 11(1):165–185, Jan. 2006.

[182] Z. Shao, Q. Zhuge, M. Liu, C. Xue, E.H.-M. Sha, and B. Xiao. Algorithm and analysis of scheduling for loops with minimum switching. *International Journal of Computational Science and Engineering (IJCSE)*, 2:88–97, 2006.

[183] Z. Shao, Q. Zhuge, C. Xue, and E.H.-M. Sha. Efficient assignment and scheduling for heterogeneous dsp systems. *IEEE Trans. on Parallel and Distributed Systems*, 16:516–525, Jun. 2005.

[184] Z. Shao, Q. Zhuge, Y. Zhang, and E.H.-M. Sha. Algorithms and analysis of scheduling for low power high performance dsp on vliw processors. *International Journal of High Performance Computing and Networking*, 1:3–16, 2004.

[185] S.M. Shatz, J.-P. Wang, and M. Goto. Task allocation for maximizing reliability of distributed computer systems. *IEEE Trans. on Computers*, 41:1156–1168, Sep. 1992.

[186] V. Shestak, J. Smith, A.A. Maciejewski, and H.J. Siegel. Stochastic robustness metric and its use for static resource allocations. *Journal of Parallel and Distributed Computing*, 68(8):1157–1173, 2008.

[187] D. Shin, J. Kim, and S. Lee. Low-energy intra-task voltage scheduling using static timing analysis. In *DAC*, pages 438–443, 2001.

[188] S. Shivle, R. Castain, H.J. Siegel, A.A. Maciejewski, T. Banka, K. Chindam, S. Dussinger, P. Pichumani, P. Satyasekaran, W. Saylor, et al. Static mapping of subtasks in a heterogeneous ad hoc grid environment. In *13th IEEE Heterogeneous Computing Workshop*, 2004.

[189] S. Shivle, H.J. Siegel, A.A. Maciejewski, P. Sugavanam, T. Banka, R. Castain, K. Chindam, et al. Static allocation of resources to communicating subtasks in a heterogeneous ad hoc grid environment. *Journal of Parallel and Distributed Computing*, 66(4):600–611, 2006.

[190] A. Sinha and A. Chandrakasan. Dynamic power management in wireless sensor networks. *IEEE Design Test Comp.*, Mar./Apr. 2001.

[191] K. Skadron, M. Stan, K. Sankaranarayanan, W. Huang, S. Velusamy, and D. Tarjan. Temperature-aware microarchitecture: Modeling and implementation. *ACM TACO*, 1(1):94–125, Mar. 2004.

[192] S. Slijepcevic and M. Potkonjak. Power efficient organization of wireless sensor networks. In *IEEE ICC, Helsinki, Finland*, 2001.

[193] B. Sotomayor, K. Keahey, and I. Foster. Combining batch execution and leasing using virtual machines. In *Proceedings of the 17th International Symposium on High Performance Distributed Computing*, pages 87–96, 2008.

[194] B. Sotomayor, R. Llorente, and I. Foster. Resource Leasing and the Art of Suspending Virtual Machines. In *11th IEEE International Conference on High Performance Computing and Communications*, pages 59–68.

[195] B. Sotomayor, R.S. Montero, I.M. Llorente, and I. Foster. Virtual infrastructure management in private and hybrid clouds. *IEEE Internet Computing*, 13(5):14–22, 2009.

[196] S. Srinivasan and N.K. Jha. Safety and reliability driven task allocation in distributed systems. *IEEE Trans. on Parallel and Distributed Systems*, 10:238–251, Mar. 1999.

[197] H.S. Stone. Multiprocessor scheduling with the aid of network flow algorithms. *IEEE Transactions on Software Engineering*, 3(1):85–93, Jan. 1977.

[198] A. Sudarsanam and S. Malik. Simultaneous reference allocation in code generation for dual data memory bank asips. *ACM TODAES*, 5(2), 2000.

[199] V. Suhendra, C. Raghavan, and T. Mitra. Integrated scratchpad memory optimization and task scheduling for MPSoC architectures. *CASES '06: Proceedings of the 2006 International Conference on Compilers, Architecture and Synthesis for Embedded Systems*, pages 401–410, 2006.

[200] H. Tan and I. Lu. Power efficient data gathering and aggregation in wireless sensor networks. *ACM SIGMOD Record, SPECIAL ISSUE: Special section on sensor network technology and sensor data management*, 4(3):66–71, 2003.

[201] T.K. Tan, A. Raghunathan, and N.K. Jha. Software architecture transformations: a new approach to low energy embedded software. In *DATE'03, Munich, Germany*, March 3–7, 2003.

[202] M.K. Tcheun, H. Yoon, and S.R. Maeng. An adaptive sequential prefetching scheme in shared-memory multiprocessors. In *Int. Conf. Parallel Process*, pages 306–313, 1997.

[203] A. Terechko, E. Le Thénaff, and H. Corporaal. Cluster assignment of global values for clustered vliw processors. *CASES '03: Proceedings*

of the 2003 International Conference on Compilers, Architecture and Synthesis for Embedded Systems, pages 32–40, 2003.

[204] Texas Instruments, Inc. *TMS320C6000 CPU and Instruction Set Reference Guide*, Dallas, TX, Oct. 2000.

[205] T. Tia, Z. Deng, M. Shankar, M. Storch, J. Sun, L. Wu, and J. Liu. Probabilistic performance guarantee for real-time tasks with varying computation times. In *Proceedings of Real-Time Technology and Applications Symposium*, pages 164–173, 1995.

[206] Y. Tian, E. Ekici, and F. Ozguner. Energy-constrained task mapping and scheduling in wireless sensor networks. In *IEEE International Conference on Mobile Adhoc and Sensor Systems Conference, 2005*, page 8, 2005.

[207] S. Tongsima, E.H.-M. Sha, C. Chantrapornchai, D. Surma, and N. Passos. Probabilistic loop scheduling for applications with uncertain execution time. *IEEE Trans. on Computers*, 49:65–80, Jan. 2000.

[208] A.W. Topol, D.C. La Tulipe Jr., and L. Shi. Three-dimensional integrated circuits. *IBM Journal of Research and Development*, 50(4/5):491–506, 2006.

[209] V. Tran, D. Liu, and B. Hummel. Component-based systems development: challenges and lessons learned. In *8th International Workshop on Software Technology and Engineering Practice (STEP '97)*, page 452, 1997.

[210] F. Vahid and T. Givargis. *Embedded System Design, A Unified Hardware/Software Introduction.* John Wiley & Sons, 2002.

[211] K.S. Vallerio and N.K. Jha. Task graph extraction for embedded system synthesis. *VLSID '03: Proceedings of the 16th International Conference on VLSI Design*, page 480, 2003.

[212] J. Valvano. *Introduction to Embedded Systems, Interfacing to the FREESCALE 9S12.* Cengage Learning, 2009.

[213] V. Venkatachalam and M. Franz. Power reduction techniques for microprocessor systems. *ACM Computing Surveys (CSUR)*, 37(3):195–237, Sep. 2005.

[214] T. Vercauteren, D. Guo, and X. Wang. Joint multiple target tracking and classification in collaborative sensor networks. *IEEE Journal on Selected Areas in Communications*, 23(4):714–723, 2005.

[215] A. Vincentelli and G. Martin. A vision for embedded software. In *CASES'01, Atlanta, Georgia, USA*, Nov. 2001.

[216] P. Vitharana. Risks and challenges of component-based software development. *Communications of the ACM*, 46(2):67–72, Aug. 2003.

[217] S. Wallace and N. Bagherzadeh. Modeled and measured instruction fetching performance for superscalar microprocessors. *IEEE Trans. Parallel Distrib. Syst.*, 9, Jun. 1998.

[218] C.-Y. Wang and K.K. Parhi. High-level synthesis using concurrent transformations, scheduling, and allocation. *IEEE Trans. on Computer-Aided Design of Integrated Circuits and Systems*, 14:274–295, Mar. 1995.

[219] C.-Y. Wang and K.K. Parhi. Resource constrained loop list scheduler for dsp algorithms. *Journal of VLSI Signal Processing*, 11:75–96, Oct./Nov. 1995.

[220] L. Wang, H.J. Siegel, V.P. Roychowdhury, and A.A. Maciejewski. Task matching and scheduling in heterogeneous computing environments using a genetic-algorithm-based approach. *Journal of Parallel and Distributed Computing*, 47(1):8–22, 1997.

[221] S. Wang and K.G. Shin. An architecture for embedded software integration using reusable compoments. In *CASE'00, San Jose, California*, November 17–19, 2000.

[222] Z. Wang and X.S. Hu. Energy-aware variable partitioning and instruction scheduling for multibank memory architectures. *ACM Transactions on Design Automation of Electronic Systems (TODAES)*, 10(2):369–388, Apr. 2005.

[223] Z. Wang, M. Kirkpatrick, and E.H.-M. Sha. Optimal two level partitioning and loop scheduling for hiding memory latency for dsp applications. In *37th ACM/IEEE Design Automat. Conf.*, pages 540–545, Jun. 2000.

[224] Z. Wang, T.W. O'Neil, and E.H.-M. Sha. Minimizing average schedule length under memory constraints by optimal partitioning and prefetching. *J. VLSI Signal Process. Syst. Signal, Image, Video Technol.*, 27:215–233, Jan. 2001.

[225] Z. Wang, E.H.-M. Sha, and Y. Wang. Partitioning and scheduling dsp applications with maximal memory access hiding. *Eurasip Journal on Applied Signal Processing*, pages 926–935, Sept. 2002.

[226] M. Weiser, B. Welch, A. Demers, and S. Shenker. Scheduling for reduced CPU energy. In *Proceedings of the 1st USENIX Conference on Operating Systems Design and Implementation*, 1994.

[227] Y. Weng and A. Doboli. Smart sensor architecture customized for image processing applications. In *IEEE Real-Time and Embedded Technology and Embedded Applications*, pages 336–403, 2004.

[228] E. Weyuker. Testing component-based software: A cautionary tale. *IEEE Software*, pages 54–59, 1998.

[229] M.E. Wolfe. *High Performance Compilers for Parallel Computing.* Redwood City, CA: Addison-Wesley, 1996.

[230] M.E. Wolfe and M.S. Lam. A loop transformation theory and an algorithm to maximize parallelism. *IEEE Trans. Parallel Distrib. Syst.*, 2, Oct. 1991.

[231] K. Wu, Y. Gao, F. Li, and Y. Xiao. Lightweight deployment-aware scheduling for wireless sensor networks. *ACM/Kluwer Mobile Networks and Applications (MONET) Special Issue on Energy Constraints and Lifetime Performance in Wireless Sensor Networks*, 2004.

[232] S. Wuytack, F. Catthoor, G.D. Jong, and H.D. Man. Minimizing the required memory bandwidth in vlsi system realizations. *IEEE Trans. on VLSI Systems*, 7(4), Dec. 1999.

[233] C. Xue, Z. Shao, M. Liu, M. Qiu, and E.H.M. Sha. Loop scheduling with complete memory latency hiding on multi-core architecture. *ICPADS '06: Proceedings of the 12th International Conference on Parallel and Distributed Systems*, pages 375–382, 2006.

[234] L. Xue, O. Ozturk, F. Li, M. Kandemir, and I. Kolcu. Dynamic partitioning of processing and memory resources in embedded mpsoc architectures. In *DATE*, 2006.

[235] Y. Yamada, J. Gyllenhall, and G. Haab. Data relocation and prefetching for programs with large data sets. In *MICRO-27*, pages 118–127, 1994.

[236] F. Yao, A. Demers, and S. Shenker. A scheduling model for reduced cpu energy. In *36th Symposium on Foundations of Computer Science (FOCS)*, pages 374–382, Milwankee, Wisconsin, Oct. 1995.

[237] F. Ye, G. Zhong, J. Cheng, S. Lu, and L. Zhang. Peas: A robust energy conserving protocol for long-lived sensor networks. In *Proceedings of the 23rd International Conference on Distributed Computing Systems (ICDCS '03)*, pages 62–71, 2003.

[238] W. Ye, N. Vijaykrishnan, M. Kandemir, and M.J. Irwin. The design and use of simple power: A cycle-accurate energy estimation tool. In *The 37th Design Automation Conference*, pages 340–345, Jun. 2000.

[239] T.Z. Yu, F. Chen, and E.H.-M. Sha. Loop scheduling algorithms for power reduction. In *Proc. of the IEEE Int. Conf. on Acoustics, Speech, and Signal Processing*, volume 5, pages 3073–3076, May 1998.

[240] Y. Yu and V.K. Prasnna. Power-aware resource allocation for independent tasks in heterogeneous real-time systems. In *ICPADS*, 2002.

[241] L.A. Zadeh. *Fuzzy Sets as a Basis for a Theory of Possibility*, volume 1. 1996.

[242] W. Zhang and T. Li. Exploring phase change memory and 3D die-stacking for power/thermal friendly, fast and durable memory architectures. In *Conf. on Parallel Architectures and Compilation Techniques (PACT)*, 2009.

[243] W. Zhang and B. Allu. Loop-based leakage control for branch predictors. In *Proceedings of the 2004 International Conference on Compilers, Architecture, and Synthesis for Embedded Systems (CASAS)*, pages 149–155, Washington, D.C., Sep. 2004.

[244] Y. Zhang, X. Hu, and D.Z. Chen. Task scheduling and voltage selection for energy minimization. In *DAC*, pages 183–188, 2002.

[245] P. Zhou, Y. Ma, Z. Li, R.P. Dick, L. Shang, H. Zhou, X. Hong, and Q. Zhou. 3D-STAF: scalable temperature and leakage aware floorplanning for three-dimensional integrated circuits. In *ACM/IEEE ICCAD*, pages 590–597, 2008.

[246] P. Zhou, B. Zhao, J. Yang, and Y. Zhang. A durable and energy efficient main memory using phase change memory technology. In *Intl. Symp. on Computer Architecture (ISCA)*, 2009.

[247] T. Zhou, X. Hu, and E.H.-M. Sha. Estimating probabilistic timing performance for real-time embedded systems. *IEEE Transactions on Very Large Scale Integration(VLSI) Systems*, 9(6):833–844, Dec. 2001.

[248] X. Zhou, J. Yang, Y. Xu, Y. Zhang, and J. Zhao. Thermal-aware task scheduling for 3D multicore processors. *IEEE TPDS*, 21(1):60–70, Jan. 2010.

[249] C. Zhu, Z. Gu, L. Shang, R.P. Dick, and R. Joseph. Three-dimensional chip-multiprocessor run-time thermal management. *IEEE Trans. on Computer-Aided Design of Integrated Circuits and Systems*, 27(8):1479–1492, Aug. 2008.

[250] Q. Zhuge, Z. Shao, and E.H.-M. Sha. Optimal code size reduction for software-pipelined loops on dsp applications. In *Proc. Intl. Conf. on Parallel Proc.*, Vancouver, Canada, pages 613–620, Aug. 2002.

[251] Q. Zhuge, B. Xiao, and E.H.-M. Sha. Code size reduction technique and implementation for software-pipelined dsp applications. *ACM Transactions on Embedded Computing Systems*, 2(4):1–24, Nov. 2003.

[252] Q. Zhuge, B. Xiao, Z. Shao, E.H.-M. Sha, and C. Chantrapornchai. Optimal code size reduction for software-pipelined and unfolded loops. In *Proc. of the 15th IEEE Int. Symp. on System Synthesis*, pages 144–149, Oct. 2002.

Index